PLUTO and
BEYOND

PLUTO and
BEYOND

A Story of Discovery, Adversity, and Ongoing Exploration

ANNE MINARD
Foreword by CAROLYN SHOEMAKER

NORTHLAND
PUBLISHING

Text © 2007 by Anne Minard
All rights reserved.

www.northlandbooks.com

Composed in the United States of America
Printed in Canada

Edited by Claudine J. Randazzo
Designed by Sunny H. Yang

First Impression 2007
ISBN 10: 0-87358-915-7
ISBN 13: 978-0-87358-915-4

07   08   09   10   4   3   2   1

Library of Congress Cataloging-in-Publication Data Pending

For my grandmothers,
Martha Bayles Kochenderfer
and Ellen Hogan Minard

Ah, but a man's reach should exceed his grasp,

or what's a heaven for?

— ROBERT RROWNING

CONTENTS

foreword

Lowell Observatory is neither the oldest nor the youngest observatory of the many established in our country since the end of the nineteenth century, but it is well known for the varied accomplishments of its astronomers. It was built on the dreams of one man, Percival Lowell, a businessman with a great enthusiasm for astronomy. It has continued into the twenty-first century because of the dreams, determination, and scientific imagination and intuition of the astronomers who followed him.

When Gene Shoemaker and I came to Flagstaff in 1963, the town was small but boasted a climate of intellectual curiosity and enthusiasm, which was important to us. Four institutions attracted our attention: the Museum of Northern Arizona with its research center; the college, not yet a university; the U.S. Naval Observatory; and especially Lowell Observatory. Astrogeologic field studies, founded by Gene, had just become a part of the U.S. Geological Survey (USGS). The Apollo program was gearing up to go to the moon. Flagstaff provided a logical home for the Branch of Astrogeology, in part, because Lowell Observatory was willing to provide use of its Clark Telescope to take some of the needed images for lunar landing sites. As the Apollo program progressed, it was difficult to obtain the necessary time elsewhere for observations. The U.S. Air Force Aeronautical Chart and Information Center (ACIC) was located on the grounds of Lowell Observatory. It was engaged in preparing lunar cartographic maps for NASA. Astrogeology and ACIC came to work closely as preparations for man to go to the moon progressed.

In 1964, a small 31-inch telescope was built on Anderson Mesa southeast of Flagstaff by the USGS for the purpose of even more lunar studies. When the missions to the moon came to an end, this telescope, close to Lowell's Perkins telescope, was turned over to the observatory with appreciation for all its cooperation. Subsequently Lowell observers used it for Mars work.

Lowell Observatory is a somewhat unusual place. At this observatory, the astronomers work on topics both within and outside our solar system. Most astronomical studies have moved to subjects beyond our solar system, with the thought that everything important here has been discovered. Lowell, however, has a long-standing tradition that says there is much yet to be found and added to the basic knowledge of our neighborhood. Percival Lowell's ambition to observe and learn about Mars and the possibilities of life there, to go on and study Venus and Jupiter, and to discover Planet X, the ninth planet, provided the foundation for many telescopic observations in the late nineteenth and early twentieth century. Founded in 1894 when Arizona was still a territory, this observatory has helped lead the way into some of the astronomy and planetary science of today. It would not be correct to say that Percival Lowell was the originator of the interest in our solar system, but he founded an observatory and wrote extensively about the things he saw. His observations of canals on Mars and features on Venus were incorrect, but they inspired the attention of other observers for many years.

It is intriguing that in 2007 we are studying Mars with even more diligence, but this time with technology unknown in Percival Lowell's time. Robotic rovers, probes, and instruments in orbit about the planet are providing huge amounts of new data. Today we talk about the possibility of manned missions to the red planet, and astronauts are as excited by that as they once were by the moon. Some of Lowell Observatory's astronomers have taken part in these later studies of

Mars. I suspect Percival Lowell would have found them a logical outcome of his own observations during a period when Mars held the same excitement as today.

Just this year, Pluto, Lowell's Planet X, has been declared by the International Astronomical Union to be a "dwarf planet." Whatever it is called, we have a space mission, New Horizons, on its way to Pluto to tell us more about this object of the Kuiper Belt. Enthusiastic interest in Pluto has remained high for more than half a century. The other planets of interest to Percival Lowell have been explored with space missions: Venus and Jupiter and Jupiter's moons have been observed with ever larger telescopes.

Anne Minard has written the observatory's history and carried her description of the astronomers who worked there and their science forward into current times. She writes from a very human point of view. It is, after all is said and done, the people who make the institution for better or worse with their varied interests and abilities, their initiative and drive, their idiosyncrasies and dreams. I have found Lowell Observatory a delightful place to work: a home of stimulating ideas held by interesting, congenial people. It is unique in having only one trustee, one who must be related to its founder. Its latest trustee saw the need and desirability to go beyond the science itself. Through his efforts, an excellent visitor's center and tours of the buildings and grounds on Mars Hill have made astronomy accessible to the public. The observatory has recognized the needs of its astronomers of today to have a telescope which is competitive and which meets their requirements without going elsewhere. Lowell is an active astronomical observatory with evolving ideas.

Carolyn S. Shoemaker
Lowell Observatory and U.S. Geological Survey
Flagstaff, Arizona

introduction

Lowell Observatory is one of Flagstaff, Arizona's most celebrated institutions, and it deserves to be. In the late 1800s, Flagstaff boasted the now-defunct Arizona Lumber and Timber Co. and a thriving population of New Mexican Hispanics lured by the prospect of working there. Basque sheepherders also made up a significant part of the population. Together, they became the founders of the town and established what are now its oldest neighborhoods. The Atlantic and Pacific Railroad first rolled through in 1881, turning Flagstaff from a struggling cow town into a source for some of the country's beef, lumber, and mineral supplies. Local mines yielded copper and uranium for decades. The 1880s also saw the arrival of two brothers from the East, David and Billy Babbitt, who would establish Flagstaff's longest-running families. Ever since, the Babbitts have helped define Flagstaff, and to some extent the state, with their business and their politics. Percival Lowell sent Andrew Ellicot Douglass out West with a telescope in the mid-1890s, to find a place to watch Mars in a favorable position. Douglass scoped out Tucson, Arizona, and Tombstone, a cowboy town farther south, before settling on Flagstaff. Lowell Observatory was established, and remains, on Mars Hill—at the top of a road that affords one of the most spectacular views of the proud and still-growing town.

Lowell Observatory has been easily as important as Flagstaff's other formative influences. Percival Lowell's early assistant, Douglass, stayed long enough to get the observatory established. But in a history that's been well documented in previous books, he left in 1901 after clashing with Lowell. Historic accounts have attributed the clash to Douglass's criticism of Lowell's belief that he saw irrigation canals on Mars. The word on Mars Hill is that Douglass was spreading rumors about an alleged affair between Lowell and his secretary, Wrexie Louise Leonard. Regardless, Douglass went on to establish several important institutions in Tucson—the now-renowned Laboratory of Tree-Ring Research and Steward Observatory. Indeed, Douglass can be credited with a founding role in the entire state's long tradition in astronomy. The 1983 book *Tree Rings and Telescopes*, by George Ernest Webb, provides an informative account of Douglass's career. William Lowell Putnam's 1994 book, *The Explorers of Mars Hill*, provides details of the observatory's earliest history. The famed discovery of Pluto at Lowell Observatory has also been well covered in literature, one example being David Levy's 1992 book, *Clyde Tombaugh: Discoverer of the Planet Pluto.*

Pluto and Beyond picks up where those treatments leave off, delving into the disproportionate contributions the small, private observatory has made over the years that have been less famous but no less important. Vesto Melvin Slipher's discovery of the redshift has been called one of the most influential discoveries of twentieth century astronomy. Art Adel, despite a short stay at Lowell that was troubled from the start, managed to make enormous contributions to the field of infrared astronomy and atmospheric research that are still in wide use today. Long-time Lowell astronomer Henry Giclas and his staff compiled the world's most widely used catalogue of nearby stars, and his long-time colleague, Otto Franz, has used every tool ever developed for the study of paired stars. And seventy-five-plus years after Pluto's discovery, several Lowell astronomers are leading the charge to study neighbors of the "dwarf planet" in the vast, unexplored region of space called the Kuiper Belt. Pluto's status change in the summer of 2006 was a good thing, they say, because it means Pluto has truly opened the door to a whole new frontier. With twenty doctoral degree-holding astronomers, today's staff at Lowell Observatory is the largest it's ever been. It is also the most productive. Lowell astronomers are using science's most advanced tools to study the cosmos, including the Spitzer and Hubble space telescopes and an army of ground-based machines. In addition the astronomers have made great use, over the years, of the two original telescopes on Mars Hill as well as four more they've built over the years at their Anderson Mesa site, leased from the Coconino National Forest southeast of Flagstaff.

It's amazing enough that the observatory, governed for more than one hundred years by trustees from a single family, has weathered the ups and downs of its own interpersonal dramas, the country's financial history—including the Great Depression—and the rapidly advancing field of astronomy. What's even more phenomenal is that Lowell has no intention of slowing down. At its one hundred-year anniversary in the 1990s, staff at Lowell Observatory made a decision to keep it a significant astronomical observatory instead of letting it become a museum. So the observatory has embarked on giant leaps to keep pace with the field. Its collaboration on the Navy Prototype Optical Interferometer on Anderson Mesa, as well as its newer partnership with Discovery Channel Communications to build the giant, highly-publicized Discovery Channel Telescope, are both incredibly ambitious and innovative projects.

The history of Lowell Observatory is about as human as any history can get. The place has been witness to struggles and strife, successes, and realized dreams. But the astronomical feats Lowell Observatory has accomplished alongside all that living—and what it's got in store for the future—are extraordinary.

Chapter 1

BEYOND PLUTO

When the International Astronomical Union voted in the summer of 2006 to demote Pluto from the realm of the planets, it caused uproar among professional astronomers and amateurs alike. The opposition took creative forms like the contest, on Jason Kottke's blog site kottke.org, to replace the age-old solar system mnemonic device "My (Mercury) Very (Venus) Elegant (Earth) Mother (Mars) Just (Jupiter) Served (Saturn) Us (Uranus) Nine (Neptune) Pizzas (Pluto)." One of the winning entries was submitted by a contributor named Brad Baxter, who wrote, "Most vexing experience, mother just served us nothing!" CNN news ran another one which, instead of just lopping Pluto off the end, accommodated the celestial bodies placed in the new category of dwarf planets: "My Very Educated Mother Just Can't

Serve Us Pizzas with Chovies X-cluded." On the positive side, the new sentence includes Ceres, which was regarded as a big, round asteroid orbiting near Jupiter before the IAU set out the new guidelines. The IAU decided that all objects big enough that their own gravity makes them round—but that aren't powerful enough to scour out the space around themselves—are to be called dwarf planets. Pluto was demoted from planet to dwarf planet. And it brings Pluto's moon Charon—newly dubbed a dwarf planet—and the dwarf planet originally called UB313 into the fold. UB313 was nicknamed Xena for some time and permanently named Eris in 2006. But the suggested replacement mnemonic is not pretty, especially for children who might not know what anchovies are, let alone their uncommon abbreviation, and who might have just learned that the word "excluded" begins with an "e." It needs work. But then, so does the IAU's definition.

For starters, the agency's definition for a dwarf planet reads, "a celestial body that (a) is in orbit around the Sun, (b) has sufficient mass for its self-gravity to overcome rigid body forces so that it assumes a hydrostatic equilibrium (nearly round) shape, (c) has not cleared the neighborhood around its orbit, and (d) is not a satellite. But Pluto and Charon, because they orbit the same center of mass, are sort of satellites of each other. And the business about clearing out the neighborhood—well, there is a whole galaxy of gray area in that. Kevin Schindler, outreach manager at Lowell Observatory, says that it is unclear how large the objects have to be that get shoved off by a true planet. "There's a lot of unnatural debris around Earth," he says. "A purist might argue that since this junk hasn't been cleared out by Earth's gravity, then Earth isn't a planet." And he said there has been no word from the IAU about how wide of a neighborhood a bona-fide planet needs to claim.

Clyde Tombaugh, who discovered Pluto more than seventy-five years ago, originally believed Pluto was larger than its currently estimated width of about fourteen hundred miles. For a while, each time astronomers took a closer look, the estimate shrank. Mark Sykes, director of Tucson's Planetary Science Institute, said years before the IAU's vote that it had become a cocktail party joke that "at the rate it's going, Pluto is going to disappear or turn into an asteroid or something," and that is how its disputed status as a planet arose.

There were deep reasons for the furor over Pluto's relegation to dwarf planet status. For decades, all school-aged kids handled models of the nine-planet solar system. We tried to wrap our heads around the idea that space was infinite, but it was acceptable to not understand that because we did understand that everything was simple, really: Earth was the third planet among nine that traveled around the Sun in an orderly, predictable fashion such as the horses on a merry-go-round. We knew our place as Earthlings because we knew our place in the cosmos—or at least our little corner of it. It was of little import to most of us that Pluto's orbit was askew compared to the rest of them. As for its size, well that made sense. It was really far away from the warmth of the Sun. It is cold out there on the fringe, and hard to grow. Pluto may not be at the edge after all, but rather just one of many objects in the Kuiper Belt, orbiting thirty to fifty times farther from the Sun than Earth. And now scientists are going even farther afield, by theorizing matter in the so-called Oort Cloud beyond the Kuiper Belt's edge. What is worse is that both regions are hurling big rocks at us, and a direct hit could mean the end of mankind. Throw in the reality that other astronomers are searching for planets, Earth-like planets, around stars far beyond even the Oort Cloud, and you understand why people are intimidated by the whole ball of wax. We are no longer riding a merry-go-round. Now we are just tiny people on a pin-prick planet, orbiting one of countless stars in a vast sea of unseen perils. Pluto is not a planet anymore. And we are very, very small.

Of all people who you might expect to protest the IAU's new definition—and Pluto's demotion—Lowell astronomers should head up the list. The observatory's claim to fame for more than seventy-five years had been Clyde Tombaugh's discovery of Pluto on February 18, 1930. If Flagstaff residents do not know anything else about Lowell Observatory, they know that fact. Cyndy Cole, a reporter for Flagstaff's daily newspaper the *Arizona Daily Sun,* said she felt the need to be gentle the day she called to ask Lowell scientists about the IAU's new definition. "I tried to ask a couple of friendly questions first," she said.

But Will Grundy, a Lowell astronomer who studies objects in Pluto's neighborhood and beyond, says Cole needn't have worried. It was important to push for exploration of "the last planet" when he and others were trying to get grant money for NASA's New Horizons mission to fly by Pluto and into the Kuiper Belt. When the mission was still in the planning stages, Grundy helped choose which instruments to send. And he fought for it when NASA almost pushed it off the priority list. But now that the mission is well on its way, the astronomy community is free to call Pluto anything it wants. Though he takes issue with the IAU's final definition, he thinks there are good points to having any definition at all for planets. It gives astronomers a starting point for talking about the objects they are just now able to see beyond Neptune. And at least it sets up a filing system for the countless objects in Pluto's region and beyond that still await discovery by Earth's ever-increasing seeing power. Today's Kuiper Belt astronomers at Lowell are happy for that, and for the fact that Pluto will be joined by countless brethren that they and their peers can study.

"The key to Pluto is that it unlocked the world of part of the outermost solar system," said astronomer David Levy, discoverer of twenty-one comets, and counting. Levy also penned a book about Clyde Tombaugh and is a frequent contributor to popular science journals. "Before 1930, we had no idea what the solar system was like beyond Neptune." Levy calls Pluto, at three billion miles from the Sun, the "king of the Kuiper Belt" because it is found in that swath of space beyond Neptune that teems with smaller rock-and-ice objects like asteroids. The discovery of Pluto opened the door to Levy finding so many of his comets, which often originate in the Kuiper Belt.

"The Kuiper Belt is so important because it actually consists of things that were used in the building of planets, unchanged since the formation of our solar system," Levy said. "If we can study those things, we can learn about our own past." And that is exactly the goal of NASA in its New Horizons mission, which is expected to reach Pluto by 2015.

· · · · · · · · ∘

About a week before the final vote, the IAU released word that its new definition for planets might affect Pluto. One option was to identify twelve planets, including Pluto, its moon Charon, the big asteroid Ceres, and Pluto's cousin UB313, a.k.a Xena. But those four bodies would be in a sub-class of planets called "Plutons." The term nearly made Grundy blush. "I don't think I'm gonna use that word," he said. The term got cast aside in the final vote, but Grundy thinks things got even worse at the IAU meeting. Some of the scientists who were there "managed to engineer the adoption of a very different definition," he laments. "In my view, the adopted definition is less scientifically useful than the proposed one was." Grundy is wary of "trying to impose categories on nature, especially this early in the game when we have studied so few planets." He admits the planets we can see right now seem easy enough to group into three categories—rocky, gas giant, and icy—but as we learn more about other planetary systems in the galaxy, "I suspect any definition tuned to our own solar system is going to start looking more and more parochial." You might say Grundy is a rebel when it comes to these kinds of definitions. For example, you will not see the term Kuiper Belt very often in his scientific papers. He calls those distant objects trans-Neptunian objects, because their orbits cross that of Neptune, now the eighth and final planet. Grundy spells out several reasons for this departure from the accepted lingo of his field. For starters, he says, Gerard Kuiper was not the first astronomer to propose a debris belt outside Neptune. Frederick Leonard did so as early as 1930, followed by Kenneth E. Edgeworth in 1943. Not until 1951 did Kuiper—who would start the Lunar and Planetary Laboratory at the University of Arizona nine years later—postulate that the belt contained materials left over from the formation of the planets, called planetesimals, and was the source of short-period comets. Some have proposed that the region should be called the Kuiper-Edgeworth Belt, out of fairness, but Grundy thinks that's pretty unwieldy. Besides, Kuiper believed that all Kuiper Belt Objects (KBOs) would be found orbiting the Sun in the same plane. A whole additional class of bodies orbits at the right distance to count as KBOs but in the wrong plane. They are scattered around the Kuiper Belt "proper," Grundy points out—possibly having been bowled off course by violent, long-ago interactions between the giant planets. Gerard Kuiper did not predict those.

The Kuiper Belt proper was named in 1992, after the discovery of 1992 QB1 by David Jewitt of the University of Hawaii and Jane Luu of MIT's Lincoln Laboratory. Since then about one thousand Kuiper Belt objects have been detected, including several worlds a third as massive as Pluto and larger. Some of these objects are very large—the biggest having a diameter of more than one thousand kilometers, or about six hundred twenty miles. Scientists now believe at least ten percent of large Kuiper Belt Objects have moons. They believe tens of thousands or even one hundred thousand icy KBOs await discovery as they slowly orbit the Sun, and they are eager to feast their eyes on them. There are purely scientific reasons: KBOs are believed to be relics of the solar nebula, the original disk of gas and dust that gave rise to everything in the solar system, and therefore should contain pristine records of the solar system's birth. There are pioneering aspects of discovering anything new and unknown. And each of the Lowell astronomers working in the Kuiper Belt has reasons to be there that are more personal.

For example, Bob Millis, Lowell Observatory's director, has spent most of his adult life on Mars Hill. In 1965, he married his wife Julia. "It was our honeymoon coming here," he says. "I carried my wife across the threshold of the apartment upstairs." When he says it he gestures toward the ceiling, to the same space where long-ago astronomer Art Adel and his wife Catharine lived in the late 1930s, and where longtime astronomer Otto Franz stayed on his first trips to Lowell in the 1950s.

Millis earned his doctorate studying Delta Scuti, a small class of variable stars, at the University of Wisconsin in Madison. And when he first arrived at Lowell, he was working for Bill Baum, the first, last, and only director of the short-lived Lowell Planet Center. The center did not stick around past the 1960s, but planetary research continues to be a theme on Mars Hill. "I spent my whole career here doing Earth-based observations of solar system objects," Millis said. "By the mid-1980s, I was getting pretty discouraged." Millis did his first studies on Jupiter's four Galilean satellites—moons discovered by Galileo Galilei—when they were points of light in a telescope. "After fly-bys, there's not a whole lot you could add from the ground," he said. "The field really seemed to be … maybe 'dying' is too strong a word." But when 1992 QB1 was discovered, a whole new world opened up, and it fell "squarely in the purview of ground-based astronomers." Jewitt discovered 1992 QB1 in images obtained with the University of Hawaii's 2.2-m telescope at Mauna Kea, and it was the first documented object orbiting in the Kuiper Belt. Soon after that, Millis became the principal investigator for the Deep Elliptical Survey, an initial reconnaissance of the Kuiper Belt conducted through the National Optical Astronomy Observatory—which maintains telescopes at Kitt Peak southeast of Tucson and the Cerro Tololo Inter-American Observatory in Chile—and the National Solar Observatory, with telescopes at New Mexico's Sacramento Peak and Kitt Peak. The collaborators aim to understand the shape and size of the Kuiper Belt and nail down the ordering of the objects that comprise it.

Millis works on the survey with Marc Buie, who arrived at Lowell Observatory in 1991 after completing a doctorate degree at the University of Arizona, a post-doctoral fellowship at the University of Hawaii, and a three-year position at the Space Telescope Science Institute in Baltimore. He had been studying Pluto since 1980, midway through his PhD program in Tucson. "I tell people it's more like Pluto picked me to study it." Buie was studying for his early comprehensive exams, the time when most students in his program were choosing their major fields of study and justifying their choices. It was a nerve-wracking time, "super high pressure. You have to stand up there and present a research project. It's not binding, but you're already halfway through grad school." He was waffling between two projects: an analysis of one element in the atmosphere of Venus, and a study of which chemicals on the surface of Jupiter's moon Io were causing its signature rainbow terrain. But his advisor had written a paper about Pluto, and Buie picked it up absentmindedly and started reading. "I couldn't put it down. A week went by. I was doing all these little calculations, just working through the physics of what it would be like to be on Pluto, physics which at the time nobody knew." He got startled out of his reverie when a fellow student poked his head in Buie's office one day and reminded him about his encroaching exams. But Buie was not just dawdling. He'd changed his mind: "I think this stuff with Pluto is just way too cool," he remembers thinking. He decided to test his professor's prediction that methane was in Pluto's atmosphere. Buie didn't think that was quite right; he thought there was ice on the surface. "Since that time, I have not been able to put Pluto down."

His first result was to prove that the methane visible on Pluto was on its surface and not part of its atmosphere. Since then he has worked on composition maps of Pluto and Charon, refinement of the orbit of Charon, and even the newly discovered satellites. He is conducting a long-term project to monitor Pluto's brightness changes on time scales a decade or longer. One of his recent, ongoing projects is aimed at producing a new generation of Pluto maps based on images taken with the Hubble Space Telescope. A noted computer whiz, he has written his own software programs for several of these projects; at times his office is a jungle of computer hard drives and parts.

Buie marvels at the quick pace of research aimed at Pluto and its companions. "The first year I was at Lowell, there was no such thing as a Kuiper Belt," he marvels. "Now I've been part of a team that's discovered about one thousand of these things." Besides analyzing the bodies with software he has designed, Buie has secured telescope time for his work on the Keck, Spitzer, and Hubble space telescopes as well as an assortment of ground-based machines. Still, he is excited about Lowell's future with the Discovery Channel Telescope now under construction. If Lowell finds a partner to fund a ten million dollar wide-field camera, the telescope could be capable of shedding light on the very structure of the Kuiper Belt. And because their access to the telescope would be largely unfettered by competition from outside, Buie's and other astronomers' projects could proceed at lightning paces. "It could lead to a picture of the structure of the outer universe," Buie says of his own work, "the composition, and what that tells you about how the solar system formed."

Already, astronomers' theories about how the region formed are dramatic. There is good reason the belt does not start until thirty times the Earth-Sun distance, or 30 AU. During the early period of the formation of the solar system, Neptune's orbit is believed to have migrated outward from the Sun due to interactions with minor bodies like asteroids. In the process, Neptune vacuumed up or gravitationally ejected all the small objects closer to the Sun than about 30 AU. But there is also an abrupt edge at 50 AU, the outer limit, and the reason for that is less clear. Astronomers think Uranus and Saturn could have traded places in the early formation of the solar system, causing a ruckus with any objects in the vicinity. And they wonder what was going on with Neptune.

They wonder if it could have slung objects from the classic Kuiper Belt out into the scattered, trans-Neptunian region. They are even tempted to blame the giant outer planets for the periods of heavy bombardment that left so many craters on the moon. "No one has even come up with a satisfactory explanation," Buie says. "It could have been stuff coming in because of the interaction of the outer planets."

Will Grundy is the third Lowell astronomer who focuses primarily on the Kuiper Belt, studying icy surfaces of outer solar system objects including Pluto, the icy satellites of giant planets, centaurs, and Kuiper Belt Objects. Centaurs are an icy class of planetoids between Jupiter and Neptune, named after the Greek class of mythological beings that are half human and half horse. Grundy is just as enthralled with Pluto as his colleagues. "In the past decade, we have progressed from thinking of Pluto as a barely resolved ... point of light to seeing it as a unique world with complicated seasonal cycles and with surface regions exhibiting diverse appearances and chemical compositions," he says. Grundy doesn't work on the Deep Elliptical Survey. He is a science team member on NASA's New Horizons mission to Pluto and the Kuiper Belt and he is already on the edge of his seat for those first results. But behind the anticipation rings a bittersweet note: New Horizons foretells the end of an era that has barely begun. Buie thinks the New Horizons mission, while a huge boon to the exploration of the outer reaches of the solar system, places some stress on the research that has been aimed at Pluto so far. "It will transform this place from an astronomer's to a geologist's world," he said. "From an astronomer's point of view there will be little left to do." Buie feels he and other Earth-based Pluto enthusiasts have nine years left, before New Horizons reaches Pluto, to "tidy up all the projects and get answers out" to support a "pre-New Horizons view of Pluto." And so in ten short years, the Kuiper Belt has gone from a place that refreshed ground-based astronomy—to yet another frontier where they will lose ground to space-based exploration.

Not to worry, some say. Astronomers have already theorized the next frontier: the Oort Cloud, an even more deeply frozen, farther edge to the solar system thought to occupy the region from 50 to 100 AU from the Sun, which is almost a quarter the distance from the Sun to Proxima Centauri, the next closest star. Like the Kuiper Belt, the region also has a hyphenated name—the Öpik-Oort Cloud—arising from its ambiguous history in the human mind. In 1932 Ernst Öpik, an astronomer from Estonia, proposed that comets originate in an orbiting cloud at the solar system's outer edge. Nearly twenty years later, Dutch astronomer Jan Hendrick Oort said such a cloud must be supplying a steady stream of comets to replace the ones that get burned up after a series of passes near the Sun. According to that reasoning, the Oort Cloud contains millions of comet nuclei, for a total mass of comets many times the mass of Earth. At the rate we are building telescopes and robotic envoys, we should know something about that any day now.

Chapter 2

SO OTHERS CAN FOLLOW

Vesto Melvin Slipher got his start at Lowell Observatory well before Clyde Tombaugh spotted Pluto, and he retired from the institution half a century ago. But much like Tombaugh's Pluto, the discoveries Slipher made during his fifty-two years at the observatory still ripple through our understanding of the cosmos. Slipher was the first astronomer to understand that most other galaxies in the universe were hurtling away from the Milky Way, using a phenomenon called redshift. That's when reflected starlight shows up skewed toward the red end of the spectrum when its source is receding. The discovery laid the groundwork for our current understanding of the expansion of the universe. What's strange is that Slipher's name isn't associated with the theory. He's most often a footnote in exciting accounts of how Edwin P. Hubble, an indisputably pompous and abrasive character, uncovered this colossal truth—by picking up where Slipher left off.

Slipher was hired at Lowell Observatory in 1901, having just graduated with an undergraduate degree from Indiana University. His brother, Earl Charles Slipher, joined Lowell five years later and became another of Lowell's best-known early astronomers. Joseph S. Tenn, a professor in the Department of Physics and Astronomy at Sonoma State University, pointed out in a talk at the 2006 meeting of the American Astronomical Society, titled "Why Did V. M. Slipher Get so Little Respect," that the seeds for Slipher's lack of fame may have been planted very early in his time at Lowell. Slipher arrived when his boss, Percival Lowell, was eliciting attention and sparking controversy by his insistence that he saw canals on Mars. Certainly individual astronomers at Lowell earned professional respect, but it took decades before Lowell Observatory's reputation fully recovered. Some modern-day astronomers maintain that the Slipher brothers and their colleagues in the first half of the twentieth century went overboard in their conservatism to counteract Lowell's flamboyance. Possibly for this reason, Slipher was part of a culture at Lowell Observatory that didn't emphasize publication of results in peer-reviewed journals, content to write up findings for the *Lowell Bulletin*, an in-house publication. Lowell's current trustee and Percival's nephew, William Putnam, thinks that Slipher was basically lazy and that his output—and under his direction, the observatory's productivity—plummeted following Percival Lowell's death in 1916. "Uncle Percy was the cattle prod that kept V. M. Slipher productive," he says.

Bob Millis, Lowell Observatory's current director, has different ideas. He thinks Slipher "showed a lot of ingenuity and stick-to-it-iveness to get those redshifts. He probably didn't get as much credit as he deserved." Millis says Slipher went as far as he could go using the equipment on hand. "Hubble had a 100-inch telescope, and Slipher had a 24-inch telescope," he said. So it was Hubble, not Slipher, who had the means to combine the discovery of redshifts with actual distances to galaxies, thereby proving the fleeing galaxies were actually that—and not just nearby, infant solar systems inside our own galaxy as many astronomers believed.

There was also the matter of the ugly financial wars that broke out between the observatory and Lowell's bizarre and reputedly selfish widow, Constance. She apparently crippled the Flagstaff part of the Lowell estate, including the trust that funds the observatory's operations, just before the Great Depression hit. Either by necessity or greed, V. M. Slipher put increasing amounts of energy into extracurricular pursuits: namely, a race to gobble more than thirty business and residential properties in what Slipher's nephew, Earl Slipher, calls "a grownup Monopoly game." And V. M. as well as his brother, E.C., were both active community members. The latter served terms in office at the city and later the state level.

None of those factors were in place when Slipher first arrived to Flagstaff. Indeed, he was given tools that were cutting edge at that time, and his drive to succeed was probably fueled by the fact that Percival Lowell had hired him only as a temporary assistant; the young man probably felt he should prove himself. Lowell Observatory's first major telescope, which arrived five years before Slipher, was a 24-inch refractor built by Alvan Clark and Sons. For more than one hundred years, it has been referred to as the "Clark Telescope." Although no longer used for science, the scope is a major highlight of observatory tours. Soon after the telescope's arrival, Percival Lowell commissioned a three-prism spectrograph for recording the light that is reflected or emitted by a celestial body, be it a planet, comet, or star.

Spectrometry got its start in 1817 with Joseph Fraunhofer, a famous telescope maker. He slid a card into the eyepiece of a telescope that was cut with a narrow slit to allow light through a prism. Within the horizontal band of colors progressing from red to blue, six hundred dark vertical lines showed up. Fraunhofer named the more prominent lines with letters, some of which are still used today. In 1859, German physicist Gustav Kirchhoff took the method a step further when he demonstrated that the lines corresponded with certain laboratory chemicals vaporized in a flame. Swedish physicist Anders Jonas Ångstrom applied these findings to the Sun a year later. For each object it's pointed toward, a spectrograph yields a spectrogram—a series of lines like piano keys that reveal the specific chemical composition of the target object.

By all accounts, it took Slipher quite a while to master the spectrograph. But he eventually used it to obtain spectra of stars, planets, comets, the eclipsed moon, aurorae, and the night sky. Among his best-known achievements, he spectroscopically confirmed the rotation period of Mars, found that Venus rotates very slowly, nailed down the rotation period of Uranus, and honed astronomers' understanding of the periods of Jupiter and Saturn. Even though he ended up reporting the absence of chlorophyll in the Martian atmosphere, Slipher also claimed to detect water and oxygen there, inadvertently perpetuating his employer's folly and drawing professional disapproval from the director of Lick Observatory, William Wallace "W.W." Campbell, and others.

In many ways, Slipher made himself indispensable at Lowell. He took care of the observatory during Percival Lowell's long absences and responded to his frequent requests, written or wired from Boston, for data and spectra on a variety of projects.

On January 29, 1909, Lowell sent one such note that charted Slipher's course for the next seven years. "Dear Mr. Slipher," he wrote: "I think it might be fruitful if you were to make spectrograms with your red plates of a conspicuous green nebula and then compare the lines as yet unidentified with known ones of the spectrum of the Major Planets. It occurs to me that the two may possibly be related." The request marked a departure from the planetary work Slipher had been doing to that point. And it roped the young astronomer into a controversy about spiral nebulae—spinning clouds of celestial dust and gas—that had been touched off more than one hundred fifty years earlier.

In a short biography of Slipher for the National Academy of Sciences, historian William Graves Hoyt pointed out that for decades astronomers had been spying on the distant, faint objects with little understanding of what they actually were. "Some believed that they were vast aggregations of stars beyond the Milky Way," he wrote, "island universes, as suggested by philosopher Immanuel Kant in 1755" and later perpetuated by Heber D. Curtis at Lick Observatory. It was this theory that would eventually win out, forcing Earthlings to re-evaluate our place. But in 1909, scientists had barely begun to entertain the idea that the Milky Way wasn't the only galaxy in the universe. Heated debates arose between Curtis's camp and that of Harlow Shapley, at Mt. Wilson Observatory in California. Shapley put a lot of stock in the work of his friend and fellow astronomer Adriaan van Maanen, who thought he had seen the nebulae moving relative to the backdrop of stars in the Milky Way, indicating they were closer than the other stars. Shapley and his followers thought the nebulae were nearby clouds of gas, within the galaxy, where new stars were likely to form.

Lowell's use of the word "green" refers to an early system

of classification that's hardly remembered, much less used, by astronomers today. At that time there were only two classes of nebulae: green and white, and the difference was in the appearance of their spectra. In general, white nebulae seemed to be emitting light from internal stars and green nebulae were believed to reflect light from some other nearby source. Less than two weeks after Lowell's first mention of the nebulae, he decided he wanted both kinds. "Dear Dr. Slipher," he wrote on February 8, 1909, "I would like to have you take with your red sensitive plates the spectrum of a white nebula—preferably one that has marked centers of condensation. Always sincerely yours, Percival Lowell." Two starred, handwritten notes at the bottom gave Slipher a little more direction: "continuous spectrum" and "but I want its outer parts." Lowell wanted Slipher to get spectra of the "outer parts" so he could compare them to spectra of planets in our own solar system. So it seems he, too, was operating from the belief that the nebulae were the birthplaces of new stars and their attendant planets.

Slipher's next letter to Lowell probably passed Lowell's in the mail. Dated the following day, it makes two requests of Lowell: Slipher wanted Lowell to send him and his brother on an expedition to South America for the Mars opposition, possibly with a helper. "Another project has occurred to me and I suggest it to you for what it is worth, but not as offsetting the S.A. expedition," he added. "It is to arrange with Professor Howe of Chamberlin Observatory of Denver for this Observatory to send an observer there for a few weeks or months work just as the conditions of seeing there warranted." So Slipher wanted to cover Mars from outposts in both the northern and southern hemispheres.

Lowell sent another letter a week later wherein certain points were actually numbered. He would use this seemingly assertive format several more times in the years ahead. "Cannot you manage to take the spectrum of the 3rd satellite of Jupiter now for the red end of its spectrum, to bring out any telluric lines it may show? I know it is difficult but I think you can bring it about," reads one of the items. Then, "Did you ever publish your spectroscopic determination of Jupiter's rotation showing how close you got to the visible one? If not I think it should go as soon as possible into a *Bulletin*." And finally, a denial of Slipher's travel request: "Your project for observations from the Peaks (from your letter of Dec. 29) commends itself to me distinctly and in this connection the more I think of an expedition to South America the less it seems to me advisable as your brother will have a lot of interesting work to do nearer home i.e. on the Peaks, Denver, etc., to say nothing of the reflector." By the Peaks, Lowell meant the San Francisco Peaks— beautiful mountains to be sure but not promising much in the way of an exciting trip, as they are fifteen miles from downtown Flagstaff.

When Slipher did write back about Lowell's nebula request, he hardly seemed to embrace the assignment. "Yours of the 16th and the 20th have been duly received and also the 8th," Slipher finally wrote toward the end of February 1909. "I shall try to answer all of these now." But before addressing the nebulae, there was news. In the following years this became a pattern in Slipher's letters; there was an uncanny amount of big news when he appeared to be behind. This time, Slipher was under quarantine because his daughter, Marcia, had scarlet fever. "I am taking precaution to disinfect my mail," he said. "This all makes operations hampered. The other men are all free of the quarantine, so it is not as bad as it might be. I have never had the fever and while the adults seldom take it, I was so thoroughly exposed to it before it was recognized that I am a little afraid. Marcia seems to be getting along very well." When it came to the prospects of carrying out Lowell's wishes, the news was grim:

> I do not see much hope in getting the spectrum of a white nebula because the high ratio of focal length to aperture of the 24-inch gives a very faint image of a nebula … This would mean 30 hours with the 24-inch for direct photograph, and as the dispersion of the spectrograph shold (sic) be at least 100-times the slit-width in order to get detail, it would seem the undertaking would have to await the [40-inch] reflector.

Slipher never did get to use the 40-inch reflector. When it arrived later that year, Lowell mounted it ten feet below ground to keep it cool. That introduced problems with turbulence, which made it less than ideal for the projects for which Lowell did initially employ it: observations of Mars and his trans-Neptunian planet search. Meanwhile, Slipher made moves toward the nebulae. He started by writing prolific letters asking the advice of spectroscopists Edward A. Fath of the Lick Observatory and Edwin B. Frost of Yerkes Observatory, and experimenting with various instrumental and photographic techniques.

And then he took a long vacation. "About yourself: you are quite right—you not only deserve a vacation but you need one; for Flagstaff is a very wearing place on the system and I therefore beg you to go off when and for as long as you please," Lowell wrote him in mid-March. Slipher wrote Lowell April 25 from Warsaw, New York, after being in Bloomington, Indiana and inquiring about a doctorate degree. He may have stayed out of touch for some time, because there's no correspondence in the archives for several months and another numbered letter arrived for him on July 19: "1. Are you a Dr. now? 2. Are the exhibits for Winnipeg getting on? Jupiter photographs etc. etc? 3. Can an enlargement of the Ganymede spectrum be made to show absence of atmospheric bands in reproduction?"

There's an unusual lull in the record of correspondence around this time. But by September 4, Lowell was after something else: "Dear Mr. Slipher: Please take spectrograms of the Sun." In January he wanted "something spectroscopic about Halley's comet and also photographs of it from Mr. Lampland ... another transparency of your brother's Saturn ... and some large sized images of Mars of this opposition and of the last ..."

The lags may mean some letters didn't make it into the archives, or they could indicate periods when Slipher was too busy to write. That seems more likely; many of his letters begin with hurried apologies for being behind. "Dear Dr. Lowell," he wrote on February 17, "Please find enclosed several tardy weekly reports." On March 7, Lowell reminded him to "write as soon as possible your acceptance of the election into the American Academy. The election lapses if not accepted. This you probably did not know. I am keeping it open as best I may." And in October, Lowell asked, "Why will you not write me a letter expressing your general disapproval of the election of Campbell to the American Academy either in particular or in general against members being elected for official positions—or both. I have already received such letter from Davidson and Burnham. If you do this please do so at once." Campbell, incidentally, was the same one who had publicly criticized Lowell for the Mars canal work.

Lowell may not have known it, but by then Slipher was making great strides toward his request of almost two years before, to get the spiral nebulae spectra. He'd learned that focal lengths, apertures, and the spectrograph's prisms weren't his problem—and he put his attention to the speed of the spectrograph's camera lens. By November 1910, he had devised a single-prism spectrograph "from equipment on hand," he wrote to Fath at Lick in late 1910. He told Lowell it "requires only about a hundredth part of the exposure required by a three-prism arrangement."

On November 9, he wrote to Lowell that he had photographed the Andromeda Nebula and it "seems to show faintly peculiarities not commented upon" by previous observers. "I shall let another plate be exposed when there is an opportunity, for the spectrum seems to be decidedly different from the solar type." By December, he was explaining to Lowell that earlier observations of the nebula were made with large reflecting telescopes, "and the idea seems to go undisputed that a long focus telescope and of course a refractor is unsuitable for such work. But I convinced myself that I knew of no reason why the focus-to-aperture ratio had the slightest part to play in spectrum work on extended objects, and this plate proves the proposition completely to my mind." Slipher was referring to the two main telescope styles. Refracting telescopes use glass lenses to focus the various wavelengths of light into useful images. Because of the physics required to do so, they tend to be about fifteen times as long as their aperture is wide. That is what Slipher meant by saying that "a long focus telescope and of course a refractor" was thought to be unsuitable. Refracting telescopes are size-limited, because the lenses can only be supported at the edge. Anything larger than the world's largest—in 1897, 40-inch (aperture) telescope at Yerkes Observatory in Wisconsin—would sag under its own weight. That scope is sixty-three feet long. Reflecting telescopes use mirrors instead of lenses to create their images. The physics are different and the focal length is only five times the width of the aperture. That translates into cheaper construction costs. Most of the world's biggest telescopes are of the latter variety.

In August 1911, Slipher was writing to tell Lowell about two small earthquakes that rattled Flagstaff, and the arrival of a new carriage. That same month, one of his letters looked a little like it had been written by someone convalescing on the grounds, not working there:

> Dear Dr, Lowell:—I am glad to say that your garden has recovered wonderfully well from the effects of the hailstorm. The zinnias had the woven wire over them and consequently they were fortunate in having very good protection from all but the smallest hail-stones. They are big fine plants now just about ready to bloom. The pumpkins though the worst hit have made good use of the roots they had left in tact and have done wonderful work in sending out new leaves and vines. Now in spite of their setback they are blooming and thus promise a crop. The gourds are coming on too. The hollyhocks are superlatively fine, their foliage being tougher and not much injured by the storm and they stand in a body higher than one's head and a more beautiful display of color could not be imagined.

The next month, Slipher took another long vacation. "Do not hasten back," Lowell wrote him on November 13, 1911, "By all means stay east over Thanksgiving and as much longer as you like." A month later, Lowell wrote to tell him to bring a low-voltage transformer, a large turkey and "a good sized goose" on his return trip to Flagstaff: "You will be met Sunday on arrival."

Then Slipher, and Lowell for that matter, seemed to get distracted for nearly a year with the mundane details of running the observatory. For almost all of 1912, Lowell's letters carried directions about how to upgrade the property and very little about getting to work at the spectrograph. At the time, Lowell was building a fence around part of the observatory land. Slipher's duty was to oversee the tree-cutting that would pave the way, and that responsibility entailed wrangling that made it all the way to Washington. "Do not of course cut the white oak, and cut the young pine with an eye to their proper thinning out," Lowell wrote him toward the tail end of June 1912. "Also be careful please that the men do not cut the few rare pinon and cedars." He wanted to cut pines instead, but he ran into trouble there because, as Interior Secretary James Wilson wrote in a letter to the attorney general, "The young pine timber which Prof. Lowell desired to cut was exactly the class of material which should be preserved in the interests of the future forest growth in this area." Nevertheless, he wrote in deference to rights that had evidently accompanied the purchase of the land, "I have been glad to instruct the forester that any restrictions upon the use of any of the material on this land by Prof. Lowell which he desires be removed."

That fall, Slipher decided to break some ground of his own. "If I can get the short focus camera lens to work properly with the three-prism spectrograph, as there is I think some slight hope of getting in that way a rough (very rough perhaps) idea of the velocity of the nebula in the sight line," he wrote to Lowell in late September of 1912: "This is very much like trying to do the impossible since others (some of them) thought they had succeeded when they got two or three lines only. But others apparently did not go about the problem in what I thought was the most promising way …" A week later he wrote, "If I succeed in getting any spectra worth while, I might try to measure them for velocity … But of course there is no rush as I do not know if it is possible to get such spectrograms." But then, he made two spectrograms of Andromeda with exposures extending over two nights, on November 15-16 and on December 3-4. These were encouraging, for the plates contained more spectral lines than had ever been recorded. "Of course the spectrum is very faint," Slipher conceded to Lowell, "and getting the velocity from the spectrograms would doubtless impress these observers as quite a hopeless undertaking, and maybe it is but I want to make the attempt."

For many reasons, Slipher could not have known that there would someday be an Internet-based phenomenon, "answers.com," and his daring spectrum of the Andromeda Nebula would land his name under one of just four major accomplishments in astronomy cited for the year 1912. And it was a big year. Two of the other noted accomplishments were happening at Harvard Observatory. A woman named Henrietta Swan Leavitt—the deaf daughter of a minister—was studying so-called Cepheid variable stars in the Small Magellanic Cloud, now known to be the nearest galaxy outside the Milky Way. She was the first to make a connection between the rate of the stars' regular pulsings and their brightnesses, which came to be called the period-luminosity relationship. Her colleague, astronomer Annie Jump Cannon, had been appointed the curator of the Harvard College Observatory the year before, and in 1912 she developed a lettered classification system for stars based on their spectra. The next year, Cannon's system got adopted by the International Solar Union and Ejnar Hertzsprung became the first to use Leavitt's Cepheid variable stars to estimate distances to stars. And more than a decade after that, Edwin Hubble would roll two of the discoveries—Slipher's and Leavitt's—into his theory of an expanding universe.

According to Hoyt's biography, at that time the radial velocities, or speeds, of about twelve hundred bright stars and a few bright planetary nebulae had been measured and all appeared to be moving slowly, at tens of kilometers per second. No one had measured radial velocities of nebulae, and no one had postulated they would be any different. But Slipher was beginning to see things in his spectrographs of nebulae that no one else had noticed. He couldn't even wait a day to tell Lowell about his spectrographs, and tipped his hand in a wire: "I hope to send by next mail a copy of a spectrogram of the nebula in the Pleiades, which means a great deal to me, because it shows so much," he wrote on December 16. In the next day's mail he sent the spectrograms and some elaboration. "The Andromeda Nebula had been done before, but not the one in the Pleiades," he wrote.

> It is the last object that is of the greatest interest, at least to me, for I feel quite convinced that this spectrogram has suggested to me the true nature of the spiral nebula. ... there seems to be no question but this nebula is shining from light borrowed from the stars in the Pleiades ... and that it is not a bank of stars so closely huddled as to give bands of light. If this is true of the nebula in the Pleiades why cannot the Spiral Nebulae be due to a central sun clouded and obscured with its own asteroidal matter and cosmical dust? I think they are just such a con fused (sic) solar system. ... without the Pleiades to light it this nebula would be dark and invisible except for the light it absorbs of the stars

beyond, and then it would appear as dark clouds against the sky like many of the peculiar dark brush marks upon the sky and Milky Way, which may have been thought by some to be due to absorbing veils, but now it seems to me that this Pleiades Nebula spectrum gives fairly direct evidence on this point. ... I must not bore you for now (for maybe I am more excited about this matter than there is cause to be).

The reply came nearly two weeks later: "Dear Dr. Slipher—Your interesting spectrograms have arrived—Why may not the continuous spectrum of the nebula be due to self-emitted light shining through cuticles of gas as well as reflected light? The Pleiades are a long way off—you know. Thank you and Mrs. Slipher for that plum pudding. We shall enjoy it to the detriment of our health." As a post script, he wrote, "I congratulate you on this fine bit of work."

On December 28, Slipher advised Lowell that he planned to get "one good carefully made spectrogram" of the Andromeda Nebula for velocity, and the following night he began this spectrogram, exposing the plate over three nights and into the pre-dawn hours of January 1, 1913. The next day he would have gotten Lowell's night letter, a sort of telegram, wherein his boss was urging him to send materials for an exhibit, including a transparency from the 1910 passing of Halley's Comet: "Please rush all things I have asked for ... I should like at once a happy new year to all. Lowell."

Despite years of Lowell's orders that were sometimes barked through overnight telegrams, the tone of Slipher's replies was always submissive and obliging (if sometimes hemming and hawing):

Your telegram was received this morning. We shall get

out the material for the exhibit as soon as possible and shall include as much as we have of suitable character for an exhibit. Since writing you before I have got another spectrogram of the Andromeda Nebula ... I feel it safe to say here that the velocity bids fair to come out unusually large. I should be able to send you something definite soon. Thanking you for your New Year's greeting ...

One has to wonder if there was any sarcasm at all in Slipher's gratitude of Lowell's "New Year's greeting." Here was Slipher, slaving away on the brink of an Earth-shattering discovery, and Lowell was barely acknowledging it in his seemingly urgent quest for presentation materials. Furthermore, the "greeting" seemed a hastily added afterthought. From that perspective, any feeling person would have been put out, even annoyed, by Percival Lowell's tone. "I'm busy, here, about to make your observatory famous!" he might have wanted to write. Instead, one gets the impression that the careful astronomer must have been biting his tongue.

Over the next two weeks, Slipher scrutinized and measured his Andromeda plates and began puzzling over an unexpected result: the nebula appeared to be moving about three times as fast as anything else in the universe. He could see this because of astronomy's version of the Doppler shift: the standard lines on the spectrogram indicating helium and hydrogen were shifted toward the blue end of the spectrum compared to where they would be if the nebula were still. The result bothered Slipher, and even raised doubts in his mind that the Doppler shift was a reliable measure of velocity. To be safe, he sent a print of his Andromeda spectrum to Edward Fath at Lick Observatory. "You will I think," he wrote, "be able to see the displacement of the nebular lines toward the violet with reference to such lines as 4325, 4308, and 4272 [Angstroms] of the Fe [iron] and V [vanadium] comparison spectrum. Other plates show the same thing, which corresponds to a velocity of 275 km ... I cannot find any other explanation." Fath confirmed his interpretation. The result was unusual in the sense that it was novel, but it also didn't match what Slipher would eventually see with most of the nebulae he observed. The majority of them would reveal redshifts, indicating they were moving away from the Milky Way. If Andromeda's blueshift had turned out to be the norm, the world would have been forced to accept that we were near the center of a universe that was collapsing.

Slipher spent another two weeks painstakingly remeasuring his Andromeda plates and found the nebula's speed to be slightly higher. On February 3, 1913, he wrote to Lowell that the Andromeda Nebula was approaching the Earth at the then unheard-of speed of three hundred kilometers per second—the value, incidentally, that is accepted today. "It looks as if you had made a great discovery," Lowell replied. "Try some other spiral nebulae for confirmation." So Slipher turned his attention to a spiral in Virgo, NGC 4594, and by April his spectrograms showed that its spectral lines were shifted far toward the red, indicating that it was going in the opposite direction—speeding away from the Earth at about one thousand kilometers per second. Slipher was astounded.

Even during this heady time, Slipher had to tend the observatory. This time, the problem came from a constant concern in the desert Southwest: water. "As I have written before, it is necessary to haul all water for the observatory uses owing to the deficiency of the city water supply which owing to its weak pressure does not bring the water up to the observatory pump house," he wrote Lowell on February 27, 1913: "The city water superintendent reports the water pressure as low as at any time and says the only relief will come from the snow melting in the foot-hills (sic) as he counts on turning water into the reservoir from the Schulz (sic) canyon." Snow or no snow, Slipher's nebular velocity work took on a life of its own for a while and began to create a stir. Percival Lowell's characteristic flamboyance, exhibited two thousand miles away, helped.

"Important details of recent discoveries at the Lowell Observatory, Flagstaff, Ariz., which tend to confirm the nebular hypothesis of the origin of solar systems, were made known here to-night (sic) on the arrival of Prof. Percival Lowell on his way to New York," reported an article in the March 23 *Chicago Tribune* and reprinted in the *New York Times* the next day. "These discoveries, according to a statement issued by Prof. Lowell, show that many nebulae shine by reflected light, and that these nebulae consist of clouds of star dust enveloped in gas. 'This is the first step in the evolution of the solar system,' says Prof. Lowell's statement. ..."

Meanwhile, the ever more cautious Slipher was still working to confirm his results. "This nebula is leaving the solar system," he pointed out to Lowell the next month, "hence it seems safe to conclude that motion in the line of sight is the real cause of these great displacements in their nebular spectra, for if there were some unknown agency akin to the pressure shifts but enormously magnified residing in the nebulae, we would not expect it to one time act one way, another time the opposite way." Slipher continued his observations through the next year. They led to his most famous moment, in August 1914 at the American Astronomical Society's seventeenth meeting at Evanston, Illinois, when he announced radial velocities for fifteen spirals. "In the great majority of cases," Slipher reported, "the nebula is receding; the largest velocities are all positive ... The striking preponderance of the positive sign indicates a general fleeing from us or the Milky Way."

The announcement brought a normally reserved crowd of scientists to their feet in a standing ovation that remains famous. It would be years—still more than a decade—before Hubble would use variable stars to confirm that the spirals were galaxies completely apart from the Milky Way. When Slipher gave his talk, and for years afterward, even he was too cautious to believe it. But on some level, every astronomer there must have understood the significance of Slipher's redshifts. Indeed, as Lowell Director Millis says in retrospect, "I believe it was the most important discovery in astronomy and possibly twentieth century science because it pointed the way toward Big Bang cosmology." It must have also been a little disconcerting. It had been three hundred years since people believed the Earth was the center of the universe, since Giordano Bruno was killed as a heretic partly for daring to suggest that there could be suns beyond our Sun, other planets beyond our solar system. People had since come to grips with that, but the suggestion that there could be autonomous galaxies fleeing from the Milky Way was, in 1914, another giant and frightening leap. "It would be comforting if we thought everything had always been this way," Millis said. But Slipher's work was the first real clue that "everything is flying apart—and it's not likely to reconvene."

• • • • • • • ○

At age thirty-nine, Slipher should have been at the top of his game, and he was indeed regarded as a major player in astronomy for that work. But, his star seemed to fade. There was a young man in the crowd at Evanston, however, who was fresh back from his education at Oxford University. In the wake of his father's death, Edwin P. Hubble had recently let go of his dad's idea that he should be a lawyer—in favor of his own lifelong passion for astronomy. He was newly outfitted with a fake English accent and the clothes of an Englishman from his time overseas, and he was starting a graduate degree in astronomy under Edwin Frost at Yerkes Observatory in Chicago. It would turn out that he was much obliged to pick up Slipher's redshifts and run with them. A little more than a decade later Hubble rolled Slipher's findings into the now commonplace— but then Earth-shattering—theory of an expanding universe.

It's not that Slipher abandoned spiral nebulae completely. That same year, he discovered evidence of interstellar dust clouds by studying the spectrum of the nebula NGC 7023. In 1930, the same year Tombaugh discovered Pluto, Lick Observatory astronomer Robert Julius Trumpler confirmed the existence of the interstellar dust by studying its dimming effect on star clusters. As one consequence, new estimates shrank the Milky Way Galaxy to a little more than half its previously believed size. Earlier estimates had assumed the dimming was a result of distance, not dust. Also in 1914, Slipher used the spectrograph to detect the rotation of a spiral nebula. By 1917, from similar observations, he concluded that these objects were all moving in the same relative direction. Their central parts, as he wrote for the *Proceedings of the American Philosophical Society*, turn into the spiral arms "as a spring turns in winding up." Around the same time, when Holland's Willem de Sitter first began to muse that the universe might be expanding, Slipher provided him with a list of twenty-five spiral nebulae and globular clusters—twice the number he revealed at the 1914 AAS meeting in Evanston. By 1917, William Putnam wrote in his book *The Explorers of Mars Hill*, only four of Slipher's velocity measures had been confirmed, but by the end of World War I, others began to take up the work and fully established their validity. In 1921, the year he was elected to the National Academy of Sciences, Slipher found that the spiral NGC 584 in Cetus was receding at about eighteen hundred kilometers per second, the fastest speed yet. Besides the American Philosophical Society's journal, Slipher published the nebular findings in several issues of *Popular Astronomy*, and of course they appeared in the proceedings from several annual AAS meetings. But at the same time, his attention seemed to get diverted very quickly after his much-celebrated finding. The energy he put into following up on his nebular work paled compared to his dedication to new topics. One puzzler is that he never did publish the nebular work in the *Astrophysical Journal*, highly respected then as now, despite numerous invitations to do so.

Edwin Frost, Hubble's mentor and at that time the editor of the *Astrophysical Journal*, may have lacked certain social niceties when it came to Slipher. In 1911, he addressed Vesto Melvin Slipher as "Mr. Victor M. Slipher." Later he commonly referred to him as Mr., not Dr., despite Slipher having earned a doctoral degree. And though his tone would get gentler in subsequent years over the same issue, he seemed almost to be admonishing Slipher when he invited him to submit the Andromeda work. "My dear Slipher," he wrote, "If you should care to write us a somewhat different account for the *Journal*, we should be glad to print it and to attempt a reproduction of the spectrum. (You know that we do not reprint from the *Bulletins* of any observatory)." Frost did encourage him to get results for more of the spirals, "for you can probably attack some more of them successfully."

By then, Slipher was well into another project. "Dear Professor Frost," he wrote in December of 1915, "The aurora I believe is frequently to be seen at your Observatory and I wish to inquire if you attempt to keep a record of its visibility. For a long time it has been a disturbing factor in the observations of the spectrum of the night sky, and, to investigate the matter, I plan to make regularly spectrum exposures for it during the next several weeks." In Frost's reply, he asked Slipher again to publish the nebula work. "Please bear in mind that we shall be glad to hear from you, for the *Proceedings of the National Academy of Sciences*, or for the *Astrophysical Journal*, accounts of your work on the rotation and radial velocities of the nebulae, when it seems best to you to contribute them to these media of publication," he wrote. Shortly afterward came another of Slipher's timely crises: "Dear Professor Frost," he wrote in early 1916, "We began last summer an office and a laboratory building which has taken a great deal of my time from my scientific work. As a result I did not get the usual amount of observing done. I hope soon to publish my work on the rotation of the Great Andromeda Nebula. It naturally would be printed in our *Bulletin*, but I should like if it might also appear in the *Astrophysical Journal* and I would be glad to send a . . . copy in case you could find room for it."

"As regards to the translation and the rotation of the Andromeda Nebula, I should particularly like to publish them in the *Astrophysical Journal*, as well as your motions of other nebulae," Frost replied a week later. But again, he reminded Slipher of the rules. "It is contrary to our practice to reprint from observatory *Bulletins*, so I would ask to have the copy somewhat different from that of the *Bulletin* itself: this would probably not be difficult for you to arrange. I should also regard it as worth while to have a short abstract or summary, covering a page or so, published in the *Proceedings of the National Academy of Sciences*." There's no evidence that Slipher ever submitted the work.

Meanwhile, that upstart young astronomer Hubble was

making headway in the nebula game. Hubble's concern with nebular classification was primarily statistical and began with his doctoral research at Yerkes Observatory around 1915. It must have been around that time when Hubble wrote an undated note to Slipher that appears in the Lowell archives: "Dear Dr. Slipher, I have been drawing up a program for investigation of nebulae, at Mr. Hale's request, to serve as a guide to systematic research work at this observatory. I shall forward you a copy of the program as soon as the draft is revised, to use in any way you see fit. I am only sorry it is not in shape to send at once." As a student, Hubble analyzed the distribution of faint nebulae primarily according to Max Wolf's 1909 classification, in which nebulae were assigned to twenty-three types on the basis of their telescopic appearance. He also found, just after he had left Yerkes for Mt. Wilson Observatory in California, a way to use Henrietta Leavitt's variable stars to establish that nebulae containing them were too far away to be within the Milky Way. Hubble's realization sparked rivalries throughout the astronomy community because it basically destroyed the long-standing positions of Curtis Shapley and his friend van Maanen.

It seems Slipher and Hubble had opposite characteristics in the realm of correspondence. Slipher was a prolific letter-writer, producing hundreds of missives that remain in the Lowell Observatory archives. But he did not publish very much. Hubble, by contrast, published well enough to achieve lasting fame—but he definitely was not celebrated for his ability to keep in touch. This became especially true when Hubble put his early career on hold to volunteer for the World War II effort. "I wrote to Mr. Hubble," Frost told Slipher about the variable nebula image, a month later, "offering to forward it to him for his examination, but have not yet heard from him. He is a Major at Camp Grant and doubtless very busy, and he is a poor correspondent. There fore (sic) I shall not send the print to him until he acknowledges my letter."

By that time it seems Slipher had all but moved on from spectral examinations of the nebula. In March of 1919 he sent Frost a paper for publication on the general auroral illumination of the night sky and the wavelength of the chief auroral line. In June of 1921, he offered the *Astrophysical Journal* a paper about the 1918 solar eclipse spectra. It was in the proof stage by the end of that year. Still, Slipher's fingerprints were visible on Hubble's early efforts. By January 10, 1918, Slipher was sending Frost "an enlarged copy of Mr. Hubble's variable nebula, which I thought would be of interest to you and him. The plate of it got a year ago is weaker than this one, but the spectrum is the same as far as one can see." In the same letter, he compared Hubble's result to his own and other astronomers' before concluding that, "the spectral resemblance is strong enough to be very suggestive of the nature of these variable nebulae."

In the years immediately following, when Hubble was gearing up to publish his theory, Slipher was heading up the International Astronomical Union's Committee on Nebulae. The IAU is an influential body that spurs observatories all over the globe to reach out into the heavens for new understandings about how they work. "Executive Secretary of the American Section of the International Astronomical Union wishes a report of our committee by March 1," Slipher wrote Hubble on February 15, 1922. "The time is becoming short for the committee report to be in and I hope you can let me have an early expression of your views on the lines of nebular work that should be carried on and the means and methods that should be followed. Very truly yours, V. M. Slipher, Chairman, Committee on the Nebulae."

Hubble's response, written just a week later, carried an odd tone. "Dear Sir," he wrote. "Your letter on the Nebulae report has just reached me. I have been considering a general nebular program for Mount Wilson and so am moved to inflict many words upon you. So many in fact I shall try to keep to simple statements, leaving the arguments pro and con to your own most competent judgment. We begin flambouyantly (sic)." He then went on to propose his detailed ideas about classification that take up three pages of solid type. That indeed got rolled into the committee's May 1922 report, about which Hubble later commented, "is most excellent and I heartily endorse it." A few months later, Hubble sent notes on a system of nebular classification for feedback from Slipher and Lampland. Slipher's belated response was characteristically apologetic, but glowing:

> Dear Hubble: Was very much interested in your communication of July 24, and it should have been acknowledged long ago. I postponed answering it thinking it would be possible to discuss it with Lampland. In addition to usual duties we have lately had to give time to eclipse plans and the arrangement of the needed apparatus and we have not got into this splendid work of yours further than to appreciate the importance of it and to admire the beautiful illustrations of the results you are securing on Mount Wilson in the nebular field. We here wish you continued success.

Slipher seemed always to be behind Hubble's vision for nebular classification, but an unpublished treatment of Hubble's expansion theory by Hoyt notes that when Hubble submitted his system to the IAU's Commission (No. 28) on Nebulae and Star Clusters in July 1925 in Cambridge, England, "it foundered on the rocks of conservatism and individualism of his contemporaries." Harvard's Harlow Shapley, who was present, might have been the one to shoot it down. After all, he had his life's work to lose if Hubble found that the spirals were indeed separate galaxies. He had been arguing for years that they were all neatly contained in the Milky Way. Shortly before the meeting, he wrote to Slipher, "I hope our designations will not be led around blindly by some prevailing hypothesis concerning the nature of nebulae." The system was indeed discussed at the meeting, though Hubble did not attend. Its members ended up deciding that the use of "terms suggestive of certain physical properties of nebulae, about which there was still much doubt, should be avoided," Hoyt wrote. "The Commission formally resolved that the classification system adopted for the new photographic catalogue be as purely descriptive as possible, and disencumbered of all terms possibly suggestive of a nature more precise than to guarantee our actual knowledge."

Slipher and Hubble continued to work together throughout this time. The tone of Slipher's letters to Hubble was always respectful, even during a long lapse in communication owing to the younger man's 1924 wedding to Grace Leib. Conspicuously absent was the bossy and somewhat condescending tone in the letters Slipher once received from Percival Lowell. And from the sounds of the letters that remain in the Lowell archives, he would have been within his rights to make demands. On August 29, he wrote to tell Hubble that their committee had decided it wanted his proposed bibliography. On October 17, apparently with no correspondence in between, Slipher wrote to ask, "How is the work of your committee progressing?" And on December 24, before which date there are still no archived responses from Hubble, Slipher wrote, "This is a request for you to send me as soon as possible your suggestions for the preliminary report of our commission on nebulae as Clusters ... Will you also be kind enough to advise what progress you and Reynolds have made with the Nebular Bibliography? ... Then again may I ask you in particular to let me have please your comments upon the four resolutions made at the Rome meeting by the Commission on Nebulae? ... Will you also be kind enough to inform me further on the progress of the several lines of work that you have been pursuing at Wilson, in their field?" Hubble replied on all counts except one on December 20. He left out any mention of the bibliography, an ambitious chronicle of nebular studies dating to 1877 that he had originally proposed. By then, he told Slipher, he was coming along in his work on Cephied variables.

"You of course know," Slipher reiterated on May 8, 1926, "that you and Reynolds and someone that you choose are expected to go ahead with your suggested bibliography." References to the bibliography eventually dropped off. But that same year, Hubble published his classification of the nebulae. Graphically depicted by his familiar "tuning fork" diagram in The Realm of the Nebulae in 1936, the scheme has been a standard feature in textbooks and other general works on astronomy ever since.

Hubble almost got scooped. Knut Lundmark, a Dutch astronomer who had reviewed Hubble's classification scheme at the IAU's 1925 meeting, published a strikingly similar one several months before Hubble could get his out. Hubble and his colleagues at Mt. Wilson would shun Lundmark for the rest of their careers because of it, but at first Hubble took it to the president of the committee: "I see that Lundmark has published a 'Preliminary Classification of Nebulae' which is practically identical with my own, except for the nomenclature," he wrote to Slipher in June of

1926. "He calmly ignored my existence and claims it as his own exclusive idea. I am calling this to your official attention because I do not propose to let him borrow the results of hard labor in this casual manner." As time has revealed, Hubble didn't really need to worry about losing credit for his work. A few years later he took another crucial step when he plotted the distances of the galaxies against their radial velocities. A clear relationship emerged: the farther the galaxy, the bigger the redshift. Double the distance and the speed doubles, triple it and it triples. This is known as Hubble's Law. Hubble never lived to see the next big step; it had to wait until more than forty years after his death. The wildcard in the equation has always been the rate of expansion of the universe, called Hubble's Constant. One of the major goals of the Hubble Space Telescope, when it launched in 1990, was to figure that out. In 1999, scientists announced they had nailed it to within ten percent of error: a galaxy appears to be moving one hundred sixty thousand miles per hour faster for every 3.3 light-years away from Earth.

But Hubble did plenty. He and his wife, Grace, went on to enjoy the fruits of fame that included months of paid vacations in their beloved England and friendships with John Barrymore, Albert Einstein, and all sorts of celebrities in between. Meanwhile, Slipher stayed on Mars Hill until his retirement in 1954—when Putnam says his father and the trustee then, Roger Putnam, had to edge him out because he was getting a bit senile. Hubble doesn't have much of a reputation for being warm and considerate; most accounts portray him as pompous and power-hungry. But where Slipher was concerned, Hubble's correspondence always retained a gentle and respectful tone. "Dear Slipher," he wrote in 1930, well into his limelight, "Would you be willing to give me your unpublished values of radial velocities of extra-galactic nebulae? Not for publication, of course, but merely to make certain that some general results we seem to be getting here are not vitiated by results on other nebulae observed at other places." On May 28, 1937, Hubble used his membership in the American Philosophical Society to endorse one of Slipher's grant requests.

And in 1953, the year Hubble would die suddenly from cerebral thrombosis and shortly before Slipher would retire, Hubble acknowledged the latter's fundamental role when he wrote to ask him to supply materials for a talk. "In May I shall give the Darwin Lecture in England, and propose to discuss the Law of Red Shifts, beginning with the initial phase and culminating in the current programs," he wrote. "Because the initial phase represented the combination of your velocities and my distances, I should very much like to show you a slide or two representing these data ... I regard such first steps as by far the most important of all. Once the field is opened, others can follow."

Chapter 3

THESE OLD WALLS

The word most commonly used to describe the atmosphere at today's Lowell Observatory is "collegial." Director Bob Millis uses it. Discovery Channel Telescope Project Manager Byron Smith uses it. Marty Hecht, the former long-time archivist, uses it. And to all appearances, it is true. There is always an air of conviviality on the hill. The astronomers conduct monthly colloquia where guest speakers are invited to give lectures. Often, a random few members of the public come to hear them. Interviews for this book were never delayed because an astronomer was too busy or important to speak. I was consistently invited to parties and games of volleyball after work. But by all accounts, that friendly and open atmosphere has evolved from something different. Long time astronomer Henry Giclas has kept a journal about his time at

Lowell in which he characterized the staff in the institution's early years as functioning under tight rein by the three older astronomers: V. M. Slipher, E. C. Slipher, and Carl Lampland. They wanted it known that they were in charge, that the younger staff members were underlings. Furthermore, they didn't want any showoffs around who would shed light on their slow pace of productivity. Both Giclas, in his reminiscences, and Art Adel, in an interview recorded before his death, thought that was the main obstacle to Adel having an easy time during his six-year stint on the Lowell staff. He was just too good: his lifetime achievements in physics, atmospheric research, and the eventual teaching of those subjects at Northern Arizona University and elsewhere are outstanding. There are few people still living who can confirm the reason why Adel was so mistreated. The fact is, Art Adel was handled by the Lowell staff in a way that was absolutely appalling, leading up to the time the older staff members employed "sordid machinations," as Giclas has called them, to shove Adel out.

Adel was born in 1908 to Orthodox Jewish parents who had immigrated to the United States with their own parents—his mother from Poland and his father from Russia. He was born in Brooklyn, New York, but the family soon relocated to Detroit, and that is where Art Adel spent most of his childhood. From early on, it seems like Adel suffered from the demeaning behavior of men older and richer than he—a disturbing trend that was balanced over the years by an equal number of people who recognized his intelligence and ability and cut him generous and timely breaks. The first trend got under way while he was ten years old, bringing in extra money for his struggling family by caddying golf carts at the Detroit Golf Club. "I remember one person in particular for whom I caddied, Horace Rackham—he was one of the original

Ford investors in Detroit," Adel told interviewer Robert Smith, in a ninety-seven-page transcript on file at the American Institute of Physics' Niels Bohr Library. "He never tipped. An immensely wealthy man, but he never tipped, and you had to knock yourself out to get a grade of fair or good. No caddy for him was ever excellent." Adel also sold newspapers on Detroit's street corners, where he no doubt toughened up and, in his own words, "acquired a second vocabulary."

Adel showed early dedication to school. He willingly attended Detroit's Cass Technical High School from 8:00 a.m. to 5:00 p.m. and cleaned out chicken coops at a local grocery store to pay for his books and clothes. Because Adel grew up in such constant close proximity to Detroit's railroad tracks, he gravitated toward engineering and figured he would apply it to locomotives. He said this idea blinded him for some time, but eventually other perspectives started opening up. On the advice of a succession of instructors he switched the focus of his studies from engineering to mathematics and then to physics, in which he eventually earned an undergraduate degree from the University of Michigan, in 1931—even though he'd completed enough coursework for a degree in math. He immediately went on to work on his Ph.D. and finished a dissertation on the structure and infrared spectrum of the carbon dioxide molecule in 1933.

Meanwhile, then-trustee Roger L. Putnam was eager to bring Lowell Observatory into the mainstream of astronomical research, this time with the focus on the infrared spectroscopic studies of the planetary atmospheres, according to Giclas's journal. And at that time the University of Michigan was heading the charge. So Putnam wrote to the chairman of the physics department, Harrison M. Randall. David M. Dennison, a noted theoretician and Adel's dissertation advisor, suggested that Adel present a paper at the June 1933 joint meeting of the American Physical Society and the Astronomical Society of the Pacific in Salt Lake City. He thought it would be a good chance for Adel to meet V.M. Slipher, and he even loaned Adel one hundred dollars in travel money—quite a gift when everyone was struggling to break free of the Great Depression. Adel was also armed with Randall's permission to forge a partnership with Lowell Observatory.

From the beginning, things were strange.

Slipher never showed up to the meeting. Instead, he'd gone to England, understandably, to present the Darwin Lecture and to receive the Gold Medal for his lifetime achievements from the Royal Astronomical Society. So Adel returned home with the experience of a conference under his belt but nothing more to show for his one hundred dollar debt. Later he learned that Slipher would be meeting Roger Putnam at the 1933 World's Fair in Chicago, where Mr. Putnam would be exhibiting products from his Springfield, Massachusetts Package Machinery Company—and where his son, David Slipher, would be exhibiting pre-cut homes. Adel found them, and they held an impromptu meeting during which, in Adel's recollection, Putnam did all the talking.

> He said ... that the university would have to pay for everything, virtually, provide the laboratory. It was to be a one-year contract. He wanted me to do the work at Michigan. They were to provide the laboratory space, all the raw materials, an instrument maker, a glass blower, and an office, and pay for the cost of publications, everything, and the Lowell Observatory would give me $500 for the year's work ... I told Mr.

Putnam, 'I've no money and I'm working as a dishwasher to keep my body and soul together ... but I still won't do what you want for $500. However, if you double it I will, $1000.' He said, 'OK.'

By the time Adel started working on the atmospheres of the gas giants, German-born astronomer Rupert Wildt had just discovered, in 1932, that absorption bands in the spectra of Jupiter and the outer planets indicated the presence of methane and ammonia. In a series of papers from 1934 to 1935, while he was in partnership with Lowell, Adel was able to show that the absorption bands were due to harmonics of the fundamental vibrations of the methane and ammonia molecules. Arthur Francis O'Donel Alexander, in his 1965 book *The Planet Uranus*, celebrated "the great triumph of Dr. A. Adel and Dr. V.M. Slipher of Lowell Observatory":

> In order to prove whether all or most of the dark absorptions in the visible spectrum of Uranus are due to methane it was necessary, either to calculate all possible harmonics of the fundamental vibrations of methane, or, better still, to photograph the methane spectrum through such depth of gas that these harmonics would register as lines or bands in the shorter wavelengths of the visible spectrum. Adel and Slipher did both and thus solved the problem completely...

Henry Norris Russel, of the famous Hertsprung-Russell diagram of star age classes, commented elsewhere that "the later proof by Adel and Slipher ... is as beautiful an application of spectroscopic theory as one could desire to see." The only trouble was, Slipher had nothing to do with the work. Giclas wrote that "Art had found it necessary to add Slipher's name to each manuscript, large or small, sent from Ann Arbor to Flagstaff approximately each month to expedite the mailing of his monthly salary installment. It was common practice at institutions during the time to include a director's name on all its manuscripts—as a way of acknowledging the author's use of resources and paid time on the research. Still, Adel's lasting astonishment at the circumstances was obvious even in his 1987 interview with Smith, when Adel was nearly eighty years old.

> Smith: I was also wondering about the authorship of the papers. I wonder what Slipher's contribution was?
> Adel: Nothing. I had to put the names of V. M. Slipher and Lampland on many of my papers.
> Smith: Because you were using in effect the spectra that Slipher had—
> Adel: But that was published.
> Smith: So there's no real reason to put his name, except for the fact of keeping paychecks coming?
> Adel: That's right. That and keeping my job. I had to do that, and neither he nor Lampland nor E.C. Slipher, none of them really knew what I was doing, had a real understanding of it.

There were rewards, however. The observatory was so pleased with Art's achievements during that first year that they extended the contract for a second year and increased the salary to twelve hundred dollars. That was about the time Adel got permission to tunnel into the sub-basement of the Randall Physics Laboratory at Michigan to set up some infrared spectrometers. He let a solar beam in through a heliostat—a simple device for tracking the movement of the Sun—which he had built and adhered to the exterior of the building's third floor. And that is how he studied the solar-telluric spectrum, or the pattern of solar radiation as it is affected by the Earth's atmosphere, in the far infrared region of the spectrum. The investigation was key in laying the foundations of modern infrared astronomy, "for one had to understand the infrared behavior of the Earth's atmosphere in order to understand the infrared spectra of extra terrestrial objects investigated through our atmosphere," Giclas wrote.

Having heard no word about working further with Lowell Observatory, Adel took the offer of a post-doctoral fellowship at the John Hopkins University for the academic year 1935-1936. It wasn't until after he'd made his decision that he found out Slipher and Putnam wanted him to come live at Lowell. It turned out to be a beneficial delay. On the professional side, Adel learned techniques at John Hopkins that he put to good use later at Lowell. And just before he left for Baltimore, he married his sweetheart of several years, Catharine Backus. It was probably a good thing the first year of their marital life was not spent at Lowell. The couple arrived to Lowell in 1936, drawn by the promise of employment, a high and dry climate perfect for atmospheric research, and a furnished apartment. At any rate, the first two promises came true. This is Adel's recollection of his arrival to Lowell and his second meeting with V. M. Slipher, in his own words: "He said, 'What, are you here already?' I was due to begin work the 1st of September and this was Sunday the 28th or 29th of August … when he finally got over the shock of my being there, he said, 'Well, I don't have any place to put you.'" Finally, Slipher decided on a room on the second floor of the administration building, where "young men"—Slipher's way of describing himself and his two senior colleagues—had planned at one time to display transparencies. So the three new arrivals, which included Catharine's sister Dorothy, followed Slipher to the room. Only Adel's words can adequately convey that moment:

> In fact, the area for the display of transparencies hadn't even been completed. Wires were hanging out and so on. The astronomers had used this room as a place to store junk their wives no longer wanted, so in this room there was an old, broken-down bedstead with rusty springs, no mattress. Are you old enough to have seen the large round oak dining room tables? Circular, very large, with spaces for additional leaves? There was one of those that had broken down, that was stored there. There was a couch, a studio couch. The upholstery, the stuffing was coming out of it. And then there was the fireplace and some fireplace tools. This was the 'furnished apartment.' Catharine held herself in check until he left, and then she burst into tears.

The couple made do the best they could—Catharine Adel soon ordered furniture from Sears Roebuck—and Adel got on with the work he came for. Adel was on fire. He made the first definitive rock salt prismatic solar-telluric spectrum, placing the great ozone band correctly at 9.57 microns. He discovered nitrous oxide in the Earth's atmosphere, and his hypothesis about the origin of the nitrous oxide in the soils and oceans of the Earth is now a generally accepted theory. It has since been shown by other researchers that nitrous oxide is nature's main mechanism for controlling the amount of atmospheric ozone. He discovered atmospheric heavy water vapor, deuterium hydroxide, which he only would have been able to do in Flagstaff's dry atmosphere. He described how water vapor affects the transmission of infrared waves through the atmosphere and made a map of the solar-telluric spectrum that included the atmospheric effects of his newly discovered elements. His most celebrated accomplishment at Lowell was the discovery of a narrow window in the Earth's atmosphere through which astronomers can detect infrared radiation for their studies. Previously, scientists thought the atmosphere was impervious to that part of the spectrum, which hampered certain observations of celestial bodies. "The window is used by astronomers today in vital studies of the nucleus of the galaxy, and the heat energy budget of the giant planets," wrote Monica Joseph in her 1973 monograph, "The contribution of Arthur Adel to astronomical infrared spectroscopy." Finally, Adel produced the first definitive emission spectrum of the moon.

Adel made all of these achievements despite working and living conditions at Lowell that continued to be less than ideal. When he arrived, no infrared equipment was available. So for a short while, V. M. Slipher let the senior astronomer work with the spectrograph on the 24-inch Clark telescope he'd used to discover the redshift. That ended shortly after Adel used the telescope to find the carbon dioxide bands in the spectrum of Venus, which astronomer Theodore Dunham and a colleague had discovered previously at Mount Wilson. Once, when Dunham was visiting in December of Adel's first year, Adel showed him the spectrum. Apparently, it was an egregious breaking of rank, and Slipher never again let Adel use the 24-inch, "or any of the other telescopes for that matter," according to Giclas' journal. But in typical Art Adel fashion, the young astronomer landed on his feet. While he waited for Stanley Sykes, Lowell's affable and skilled instrument maker, to complete the tools he needed for his experiments, Adel scratched out handwritten calculations to describe the theoretical characteristics of the carbon dioxide Dunham had seen—and Adel had confirmed—in Venus's atmosphere. "The result was a highly specialized technical paper that no one here understood, but because Bernice (Giclas) had taken several chemistry courses, V.M. gave it to her to read," Giclas wrote. "Finally, to assuage V.M.'s fears, the manuscript was sent to Professor Dennison for review, and he recommended it for publication."

While Art and Catharine Adel had done their best with their little apartment, it started to get cold in there. Adel told Smith, his interviewer, that V. M. Slipher and his wife, Emma, were headed by train to Indiana "and I suppose Emma must have asked him, between Flagstaff and Albuquerque, what we were going to do to keep warm. It was the bitterest October I've known in Flagstaff." Slipher didn't call, which still seemed to rankle Adel all those years later, but instead he mailed a post card to his brother, E. C. Slipher, at the Albuquerque train depot and asked him to take Art Adel shopping for a stove "like the one in the kitchen," Adel recalled.

> I knew then that there would be trouble. Such a stove would be totally incapable of heating that enormous room, and I didn't want another stove like that. So I went down to Waldhouse's with E. C. Slipher and sure enough, there was a stove down there, second hand, just like the sheep herder's stove we had in the kitchen. 'Now, this is the kind of stove my brother wants you to have,' and I said, 'Nothing doing.' Equipment was being built for me by Stanley Sykes, the instrument maker in the shop at that time, and we'd invested a lot of time here, but I was ready to pick up and leave.

So Adel vehemently stood his ground and ended up with a ten dollar stove, over repeated complaints from E. C. Slipher that his brother would not be pleased. But Art and Catharine Adel spent their first winter "warm as toast," he said. They did have to be inventive in other ways, Giclas pointed out. There was normally no hot water in the main building where the Adels' apartment was when the big furnace in the basement was not going, he wrote. In the observatory's earlier days, Giclas and some of his young colleagues had heated water next to the furnace downstairs and showered there. "To heat water hot enough to get to the second floor was almost impossible," he wrote. "The Adels solved the problem by heating water in a copper wash boiler on the sheep herder's stove ... and rolling it down the hall to the bathroom. It took so long to heat the water for one bath that they both used the same bath water, Cappy first, then Art."

Adel had come from a vibrant academic institution at Michigan, and the climate on Mars Hill also took some getting used to in that regard. Giclas recalls that all of the younger astronomers—including Giclas, Adel, Clyde Tombaugh and James Edson—missed the seminars and colloquia they'd left behind in their graduate institutions. So they made their own. They'd get together for "astronomical bull sessions," as they called them, and eventually they decided to formalize the proceedings by taking turns to give talks about their respective areas of interest. They even invited the older staff members. "Adel gave the first one on some phase of infrared molecular structure. Edson gave the next one on rockets—Robert Goddard was just beginning his early rocket experiments in New Mexico," he wrote. "Needless to say the older men were not in the least interested in this application and believed ... that the Earth's gravity could never be overcome by any conceivable force. I did the third one on electron optics." Tombaugh gave his talk on the geology of the moon. Years later, Giclas read an entry in Lampland's diary from April 6, 1937, shortly after his own presentation: "Giclas gave his colloquium talk 'Electron Optics' and it was phantasy (sic)." The young men, fairly discouraged, retreated to their astronomical bull sessions "where we even discussed how nice it would be if we had some progressive and inspiring leadership instead of the attitude that, 'young people should not be seen or heard.'"

Adel did get the instruments he needed from Sykes' shop. East basement rooms were prepared for his infrared lab, and a heliostat was built for the attic roof above it to bring solar radiation into the laboratory. Sykes built the prism spectrometer under Art's supervision, and the maintenance man constructed the housing for the heliostat on the roof. With the prism spectrometer, Adel quickly completed his definitive portrayal of the rock salt prismatic solar spectrum and recorded the thermal emission of the moon, which showed it acts like a black body—meaning it absorbs all electromagnetic radiation that falls on it. None passes through it, and none is reflected. Adel was also eager to search for an atmospheric window through which astronomers might be able to see the thermal radiation of the giant planets and the nuclei of galaxies. But he would need a prism made from potassium bromide, and the older astronomers refused to buy it for him. That's when Charles Greeley Abbott, secretary for the Smithsonian Institution and one of Adel's admirers, stepped in. Abbott had made a trip to Germany and brought back two cylinders of potassium bromide. If Adel would send him the dimensions, he'd cut it into the required prism, he said. And so Adel got the tool he needed and discovered the atmospheric window, which has helped astronomers study galaxies and the giant planets ever since.

Giclas and Adel were both pretty certain it was Adel's productivity that made him a *persona non grata* with the older astronomers. Roger Putnam confirmed that Adel wasn't wanted by his colleagues when he visited one year, telling Adel that he'd put him there over the heads of the older men. And finally, when the war came in 1942, the older men got their way.

At the outbreak of World War II in December 1941, E.C. Slipher, who was on the local draft board, threatened to have Art drafted unless he left immediately for the war effort. Adel pointed out that he could secure war effort contracts and pursue the work for them right there on Mars Hill. He was turned down. In consequence of these "sordid machinations," as Giclas called them, Adel left the observatory just after the first of the year in 1942. He and Catharine drove to temporary Navy facilities in Washington, D.C., where he helped develop technologies to protect warships against magnetic mines.

Art Adel wasn't completely incapable of standing up for himself. Also around the time the war broke out, he took famous cosmologist Carl Sagan to task for attributing his discovery of the infrared window to someone else. He demanded an apology in writing, but when Sagan ignored his admonishments he took the matter all the way up through the chain of figureheads in the field. "I wrote the president of Cornell University to tell him that I didn't think he should appreciate having on his faculty someone who was no more interested in telling the truth than Sagan," Adel said. He didn't quit until he'd communicated with the president of the American Physical Society, a Chien-Shiung Wu, the first woman to be elected president of that body. She assembled a group of astronomers and physicists who pressured Sagan to publish a correction, which he finally did in *Physics Today*. So Adel didn't always willingly suffer at the hands of others—but one wonders why, after the war, he would knock on Lowell's door again with the expectation that he might return. He and Catharine had even left all their things there in early 1942, including Catharine's piano. "I felt I was on a mission," Adel would explain, years later. "I had a mission, you know, at the Lowell Observatory. It wasn't anything that they asked me to do. These projects were my own devising."

The prospect actually looked good at first. In the spring of 1946 he paid a visit to Roger Putnam and his wife in Washington, D.C. They had a great time, Adel recalled. Putnam flew Adel around the Washington area and dined him. He even made suggestions that Adel might be poised to be director of the observatory someday, something Adel said he had never even considered. But the next day, the visit ended abruptly for reasons Adel could only guess. It happened while Putnam, Adel, and others—including a friend of Putnam's by the name of Trask, who sat in the front seat while Putnam drove—were in the car together. "I sat in the back and I was very tired and I wasn't saying much, but I heard the conversation pretty much that was going on between them," Adel said.

> Trask whispered something to Putnam, and it was as though Putnam had had an ejection seat, the kind that pilots have in the aircraft. I thought he'd leave the roadster, leave the coupe or whatever it was. He didn't but he was in a state of shock. And I just had the feeling somehow it had something to do with me...

Nothing more was said, but within a week a letter arrived for Adel in Ann Arbor, where he and Catharine were visiting family.

Adel told his interviewer it contained a short note to the effect that Adel's work was better suited to the university laboratory, not the observatory. Putnam had included a one hundred dollar check, presumably to soften the blow. Years later, when Adel was unable to locate the letter, he tried, unsuccessfully, to secure a copy even when he asked Michael Putnam for it. Michael Putnam was the trustee between Roger Putnam, his father, and Bill Putnam, his brother, the current trustee. The letter didn't turn up then, and it still doesn't appear in the Lowell archives. All that exists there now is a short response written by Adel to Putnam on May 18, 1946: "I deeply regret that the Lowell Observatory does not desire my return," it reads, going on to mention that he would be leaving the University of Michigan and taking a position at McDonald Observatory in Austin, Texas. Adel would later call it an "abortive" move—it only lasted a night. Once again, the couple had been promised a place to live that turned out to be disappointing—and this time, they didn't stick it out. "Someone in the last few years authored a book on the history of the McDonald Observatory and mentioned in it that the reason the Adels did not accept the position there was that Cappy Adel found the place too isolated and lonesome," Giclas wrote. "I had to correct them on this point." Apparently, Catharine told Giclas that the couple was disgusted and insulted that the house in which they were to live—and that had been represented to them as being ready for occupancy—was a wreck: "Broken windows, paint on the walls scratched off, and dirt with decayed food (jelly sandwiches) smeared on them by the small children," Giclas reported. "The Adels felt that if the resident maintenance was that negligent and irresponsible, there would be big troubles with trying to live there. The moving van came to Lowell for the Adels' furniture about a week later, and instead of going to Texas, it was sent to Ann Arbor, Michigan."

Adel spent the next two years teaching astronomy at the University of Michigan and as an assistant professor of astronomy at the McMath-Hulburt Solar Observatory. Two years later, Adel refused an offer from the Holloman Air Force Base in Alamagordo, New Mexico to become superintendent of research at the salary of $10,305 a year, and in response the Air Force Cambridge Research Center volunteered to send a research contract to him wherever he decided to go. He took it back to Flagstaff, but this time he stayed off Mars Hill. Adel accepted an offer for $4,000 a year at Arizona State College, which is now Northern Arizona University. President Lucy Eastburn of the college agreed that Art might do research above and beyond his academic load as long as he didn't have to seek support or solicit funds within the state of Arizona. Soon enough, the Air Force contract showed up, and Adel established the Atmospheric Research Observatory on campus where he taught student observers and completed more influential work of his own. Among other accomplishments, Adel determined vertical atmospheric ozone distributions and discovered ten- and eighteen-day temperature fluctuations in the stratosphere, with a corresponding eighteen-day period in the troposphere. He also studied the amounts of atmospheric moisture with the potential to fall as precipitation above Flagstaff and in 1973, used the results to determine the imminence of Flagstaff's summer monsoons which, he noted, begin when the water in a column of air over Flagstaff reaches just under an inch. He was a professor of physics, taught mathematics, and introduced astronomy into the curriculum. "Not only had Art distinguished himself with a third career in atmospheric research," Giclas writes, "but he was a most successful, though a demanding teacher, who inspired many students to go on with graduate training and to contribute in their own fields of research." In a university-issued press release commemorating Adel's 1976 retirement, he was credited with starting "the tradition of research at ASC-NAU, a practice now rich and varied." In recognition of that and other services to the university, he was honored in the spring of 1984 by having NAU's mathematical sciences building named for him.

Adel was given many distinctions as a result of his life's work. In 1969, he received the "Achievement in Physics" award from the National Executive Council of Sigma Pi Sigma. He was elected a fellow to the Explorers Club in 1977 and the Johns Hopkins University Society of Scholars in 1978. He was a fellow of the American Physical Society and the Arizona Academy of Science, a member of Phi Beta Kappa, the American Astronomical Society, the American Association for the Advancement of Science and the American Meteorological Society. In 1946, he became the first NAU faculty member to be listed in *Who's Who in America*. His colleagues selected him to present the first annual NAU Honors Convocation address in 1959, and the students dedicated their 1964 yearbook, *La Cuesta*, to him.

"In retirement Art attended many of the numerous scientific lectures held in the Flagstaff community," wrote Richard L. Walker of the U. S. Naval Observatory in Flagstaff, in an obituary that appeared in the May 1995 issue of *Physics Today*. "Visitors were always impressed with the questions of the dapper gentleman with impeccable manners. Art's persistent, probing mind was a stimulation to his students, colleagues and vast number of friends. We miss him." Art Adel died of cancer on September 13, 1994.

As for Catharine Adel, she "lived a rich and exciting life" for nearly ten years after her husband died, said Frank Scott, coordinator of piano studies at Northern Arizona University. In the spring of 1999, Scott performed the Grieg *Piano Concerto* in honor of the widowed Mrs. Adel, whom he calls "NAU's musical benefactress."

She was named Flagstaff's 1999 Citizen of the Year by the *Arizona Daily Sun*, the local newspaper, in recognition of many donations to music education at NAU and in the community. She created in 1997 the Catharine B. Adel Music Foundation, which has provided several musical opportunities for young musicians in the community and pays for NAU student ensembles to visit local schools. She also provided scholarships for students to attend the Suzuki Institute on the NAU campus, started an annual Flagstaff spring piano festival and competition for local students, donated to NAU four pianos (including her own grand piano that's now in NAU's Ardrey Auditorium), and bought another grand piano for Flagstaff Arts and Leadership Academy, a charter high school focusing on the arts. "Her mission was to put an instrument in the hands of every child in Flagstaff," said John Burton, dean of NAU's School of Performing Arts. "And she was well on her way to doing it." Catharine Adel died in 2002 at The Peaks, a pleasant retirement community nestled in the pines in Flagstaff, where she spent the final two years of her life.

Chapter 4

STARS THAT MOVE AND FADE

I kept expecting ninety-four-year-old Henry Giclas to get tired of my questions about his fifty-plus years at Lowell. So it was sweet relief when I could get him to laugh.

"I was reading some of Luyten's papers, you know, Willem Luyten," I said, prepared for his face to darken at the mention of his former rival. "Was he crazy?"

Henry laughed. "Luyten was a crusty old guy. He'd crossed purposes with most astronomers," he began. Giclas navigated quite a few challenging personalities during his time at the observatory, where he put together his series of ten "Proper Motion Catalogs." They are the world's most frequently cited catalogs of stars close enough to move as viewed from Earth. Luyten was a professional thorn in his side, and much of their rivalry was rooted in the mutual conviction that the other got too much credit for his part in their race to map nearby stars in all the heavens. Giclas's catalogs were published faithfully from 1957 to the late 1970s, rising above Luyten's loud protests as well as sometimes odd, in-house dynamics and the tragic mental collapse of Robert Burnham after he had worked on the project for twenty years.

It was the third or fourth time I had visited Henry Giclas at his home, a brown, two-story affair in the heart of downtown Flagstaff's oldest residential section. He had said that I was welcome to visit him any time, except when he met his friends for coffee every morning at a nearby bar and restaurant called Charlie's, and except when he was eating lunch every day at the same place.

Giclas always seemed to be wearing dress slacks and socks to match, a button-down shirt, and suspenders. His hair was sparse and white and his face much thinner than in his photos from even five years earlier. I would find him sitting in his straight back leather recliner next to the phone, his basket of mail, and the calendar where he wrote down his appointments. He was always in the company of a quiet nurse, who either sat down to listen to our talks or stayed tuned with one ear while she straightened, dusted, and mopped. The television was directly ahead of his chair, but I never found it turned on.

The astronomer spent much effort chronicling his impressions and reminiscences, as he called them, from his time at Lowell Observatory. Kevin Schindler, Lowell's public relations manager, had photocopied and assembled years of the Giclas journals—about one hundred fifty pages—in a three-ring notebook and lauded them as an unmatched resource for the history of the place. The typed notes are painstakingly detailed and quite often witty, shedding laugh-out-loud light on five decades of science, professional interactions, and even a little debauchery among the Lowell staff. I asked Giclas once why he had never published his writings, and he shrugged his shoulders, pointing out that they had been incorporated into several publications on the Hill. His memory of those years never really faded, and he was always happy to travel lucidly back in time to any of the decades between the 1930s and 1980s when he was a full-time employee at Lowell. But I am reasonably sure, despite a half dozen such visits, that he never did commit my name to memory. And he may very well have received my reason for interviewing him—this book—as new information every time I explained it.

Giclas remembers that he first visited Lowell with his dad, Eli Giclas, when he was sixteen or seventeen. "I think I was more frightened about walking up through the woods on that hill at night than I was excited about looking through the telescopes," he said. His dad had arrived to Flagstaff on the Santa Fe Railroad and fell in love with the town in the summertime, then became chief engineer for the Arizona Lumber and Timber Co. He helped erect the original 42-inch refractor in 1909 and 1910. The younger Giclas grew up with the children of Flagstaff's best-known families, and their stories cling to him like campfire smoke. Just one example is his recollection of playing in asbestos piles with the Babbitt boys outside the lumber mill—well before anyone knew that was dangerous. He also went to school with David Slipher, V.M. Slipher's son, and recalls very clearly that his friend showed no desire to be an astronomer. David Slipher turned out to be a businessman. But Henry Giclas got hooked on the observatory almost from the word go. He, David Slipher, and a friend of theirs sometimes printed pictures of scenery and their girlfriends in the observatory's dark rooms, and while they were drying he'd sneak off to read books in the library.

"Lampland would point out interesting books for me to read and I'd go up after school and read them," he said. By the summer of 1929, just before he enrolled at the University of California at Los Angeles, Giclas spent all the time he could on Mars Hill. Once he got to California, he enjoyed lectures at the California Institute of Technology by a number of astronomers from Mt. Wilson. He even heard a talk by Albert Einstein, who spent time there as a resident scholar. Though Giclas doesn't mention it in his journal, his time in California also coincided with the height of Edwin Hubble's fame, though that was around the time Hubble's bosses at Mt. Wilson were complaining that he neglected lecturing close to home in favor of more prestigious opportunities elsewhere. Giclas makes no mention of attending a talk by that famous astronomer, and chances are he never did. In February of his first year away, Giclas heard the announcement that Clyde Tombaugh had discovered Pluto, and it made him nervous: "I felt the place would become so famous now that I would never be welcome there again."

That didn't happen. Instead, a number of events lined up to get Giclas an assistant position under V.M. Slipher in the early 1930s—after he'd tired of California and transferred to the University of Arizona in Tucson to finish an engineering degree. Engineers were a dime a dozen during the Depression, Giclas remembers, so he was happy to accept a sixty-dollar monthly salary with room and board. One of his first assignments was to help on a Harvard-Cornell expedition to study incoming meteors with telescopes set up at Mars Hill and Bellemont. He also helped Lampland with some of his plates of Eros—a near-Earth asteroid expected to come within ninety-three million miles of Earth in 2012—and his program to pin down the position of the newly-discovered Pluto.

February Burnham had found another 9th magnitude comet on his own 12 days after his arrival."

To look for stars that were showing movement, the researchers used Tombaugh's old photographic plates as well as images they made themselves. Sometimes the observers took turns, so one was always making news plates and the other was comparing the old and new plates with an instrument called a blink comparator. Rarely used these days because Kodak stopped making plates for them decades ago, blink comparators would permit switching rapidly from one image to the other, looking for objects that appeared to jump back and forth between two positions. It is so tedious that, in the late 1920s, the Slipher brothers handed the task off to their underling, Clyde Tombaugh—and that's why he ended up with the credit for discovering Pluto. It took a patient soul to not mind the work, and decades after the proper motion survey one such person, Carolyn Shoemaker, would use it for her small team, which included her husband Gene, and subsequently discover loads of new comets.

Giclas learned early in his project that it is best to have two people blinking the plates—and then comparing them—in order to catch the most stars. So that fall, he traveled to the University of California's Lick Observatory and UC-Berkeley, searching for a second assistant. He found Norman Thomas, who struck him as a good candidate. Thomas recalls that Giclas actually came looking for him at a meeting of the American Astronomical Society in San Francisco. "He said he'd have to start me at the bottom of the scale and all that," he said. "Henry watched his money at Lowell." Thomas agreed to start at Lowell as soon as he finished his final semester, and the original student assistant, Charles Slaughter, would stay on just that long and then transfer from Northern Arizona University to the University of Arizona's astronomy program. Thomas recalls that he bought and shored up a 1959 International truck and hauled all the possessions that he and his wife owned in a trailer behind it. The truck was so weak that it could not even haul the trailer up the road to Mars Hill. He and his wife Maryanna had to make several trips to move into their apartment at the observatory.

Toward the end of 1958, Slaughter, Burnham, and Giclas resurrected the *Lowell Bulletin*, which had been dead for more than twenty years, in order to publish a one hundred sixteen-page first installment of the proper motion catalog. Many of the pages were packed with one-inch "finding charts" that showed each star's position in the night sky. They had queried peer-reviewed scientific publications with little success. "The *Astronomical Journal* was sympathetic, but the cost was so great and at the time there was a six-month delay on publication," Giclas wrote. "Reactivating the *Lowell Bulletin* seemed to be the best way." Before the year was out, on December 30, Burnham found another comet on one of the proper motion plates. He found two more on January 21 and 26. "It always fell to my lot to measure accurate positions and send them in with the announcement," Giclas wrote. Such discoveries have long been reported to the Minor Planet Center, a sort of clearinghouse for news and discoveries about the heavens.

Giclas, Thomas, and Burnham rolled along on the proper motion surveys for two decades, churning out *Bulletin* after *Bulletin* with the exhaustive lists. The early ones were all titled the same way: "Lowell Proper Motions; Proper Motion Survey of the Northern Hemisphere with the 13-inch Photographic Telescope of the Lowell Observatory." The first catalog notes eight hundred fifty-six stars moving a quarter-second of arc—a tiny, astronomical measure of motion—per year as viewed from Earth. The closer the stars, the more they appear to move. Over the years the Flagstaff observers would get savvy enough to bump the lower limit of movement down a little, but not much. It would take the more powerful telescopes of a different observatory—and a future era—to capture the smaller proper motions of more distant stars.

But for the stars they could see, Burnham and Thomas made a winning team. The two spurred each other on to discover the most objects they could in their plates. It turned into a friendly competition, Thomas says in retrospect: "I think Bob was better," he said. "He could concentrate a little better." By the time part II of the catalog came out a couple of years later, "A total of 4154 stars have been measured at least once each for motion by two different observers," the authors wrote. "From these measures, 2272 stars with motions ≥ 0.27" a year, the working limit adopted as the most reasonable, have been retained for publication in this list. If duplicate observations from overlapping plates were combined, the number of different stars in the list would be reduced to 1792 of which 1096 or sixty-one percent are, to the best of our knowledge, newly discovered ones." Part III lists 1,152 new stars. In 1962, the trio departed from the systematic mapping to delve in detail into the Hyades star cluster. At one hundred fifty-one light-years away in the constellation Taurus, Hyades is one of the nearest star clusters to Earth and visible with the naked eye. They lowered their proper motion limit to a fifth of an arc second a year to get fainter stars, and found eighty-two of those. Overall, inside the first five years the group had mapped 8,422 stars, of which 3,421 were newly measured, as far as they knew.

By then, whether he knew it or not, Giclas was headed into a rivalry. Luyten, a Dutch-American born in Indonesia, worked at Lick Observatory and Harvard College Observatory and taught at the University of Minnesota. He'd been studying stellar proper motions since the late 1920s, and an enduring reputation as a prickly fellow might explain why he always worked alone. But it was baffling to astronomers then and now how he was able to be so productive, recording the motions of more than fifteen thousand stars by the time his work was done. Giclas was pretty clear about how he did it. "He copied about five hundred of my stars and put his numbers on them and never gave any credit to Lowell," he said, decades later. Luyten's cantankerous nature is stamped all over even his earliest papers, which routinely point out the flaws in other research in order to poise his own as being more valid. "Where the theoretical work of Russel and Walter leads to contradictory conclusions it would appear necessary first to obtain as much observational evidence as possible to serve as a starting point for future theory," he wrote in the introduction to his 1935 paper, "On the Apsidal Motion in Binary Stars." Apsidal refers to the motion of an orbiting object when it's either closest or farthest from the object it is orbiting. He goes on: "A great deal of observing material is available, but for the most part this is not in a form ready for use. In many spectroscopic orbits calculated in the early days the elements are given with entirely unwarranted precision ... due to an overdose in decimals ... For these reasons it has seemed advisable to re-examine all the available material."

By the early 1960s, the first few Giclas catalogs had been published in the *Lowell Bulletin* and presumably sent to the Minneapolis library as well as libraries at other astronomy institutions. But in Luyten's mind, he held title in the proper motion search. "Our present knowledge of the faint end of the Luminosity Function rests almost entirely on the data obtained in the Bruce Proper Motion Survey of the Southern Hemisphere," he noted in his own catalog, "Proper Motions of Faint Stars." In 1967, he published an outright attack of the Lowell program called "A Comparison Between the Bruce, Palomar-Schmidt and Lowell Proper Motions." In it, he found "no systematic differences" between the proper motions from his own survey and the Palomar-Schmidt work. Not so for Lowell Observatory. He wrote that motions from the Lowell survey were exaggerated compared to the ones that he had found:

> ...motions as large as 0.545" in the Lowell surveys correspond to 0.50" in the Bruce or Palomar-Schmidt surveys. In turn this implies that numbers of stars found with motions larger than 0.5"—which appears to be accepted as the mythical lower limit of "large" motions—in the Lowell Survey must be multiplied by 0.78 to correspond to numbers in my surveys. It is immaterial here which of the surveys is ultimately proved to be systematically correct for only the relative numbers count.

Luyten's point was clear: it was two against one, and Giclas was the odd man out. His proper motions must be wrong. In 1962, Luyten issued a catalog where he drew on, and cited, the Lowell proper motions with no apparent snide remarks. A year later, he seemed reasonably professional again when he compared the two surveys. "The only other large survey undertaken since the Bruce Survey is that now under way at the Lowell Observatory which is expected to provide data for the entire northern Hemisphere and a portion of the Southern Hemisphere," he wrote, "... the Lowell Survey is mainly an addition to the Bruce Survey in area but cannot really be said to extend it to fainter magnitudes."

Norm Thomas, who worked with Giclas on most of the proper motion survey, says Luyten's grudge against Lowell Observatory may have reached back to the days when Giclas was just drawing up proposals for the program. Around that time, Luyten wrote the then-director of the observatory, E.C. Slipher, and requested to take all of Clyde Tombaugh's plates with him to Minnesota. The director naturally refused. "After that, Willem Luyten was down on Lowell," Thomas said.

By June 15, 1965, Luyten was reporting proper motions for 11,117 stars, including more than 10,000 new ones. If there was a race on, the Lowell team had to be sweating it. But it was Luyten who cracked. A note inserted into a 1966 University of Minnesota paper brims with ego:

> After the manuscript had been completely typed, and just before it was sent to the printer, Royal Observatory *Bulletin* No. 108 was received. This is a remarkable document. It deals with the motions of stars in the region of the Hyades and purports to give copious references to previous work. My own interest in this field goes back nearly 40 years ... and since that time I have published at least eight papers dealing with motions in the region of the Hyades—but none of these are mentioned ... the values given in *Bulletin* 108 are virtually identical with mine, and their values could, at best, be considered as confirmations of mine. Giclas' survey of the Hyades is described as 'the most complete survey for faint stars in the Hyades'; ... I published the results of blinking five pairs of the Palomar Schmidt Survey plates in the region of the Hyades ... but no mention is made of this at all.

Luyten didn't slow down; in July 1966 his paper came out with fifteen thousand stars. Nor did he calm down. The complaint against Giclas was nothing compared to a lengthy 1969 diatribe against another researcher, William van Altena, now of Yale University, whose catalog of seven hundred eighty-eight proper motions of stars in the Hyades region had been published three years earlier. Luyten's tone seems hysterical, an effect exaggerated by words he underlined for emphasis. At one point the flow of paragraphs is interrupted where Luyten inserted a single, centered sentence in all capital letters: THESE OBJECTS DO NOT EXIST.

Giclas says Luyten's attitude was well known throughout the astronomy community at the time. "We'd meet at astronomy meetings," Giclas recalls. "He'd get up and start giving somebody a hard time. The chairman always knew it. He'd tell him to shut up and sit down." Furthermore, Luyten's vast catalogs of proper motion stars were essentially summaries of what Giclas and others had done in addition to Luyten; that is, they were partly plagiarized.

Astronomers came to rely on the Lowell catalogs, not Luyten's, for subsequent work with close stars. Thomas suspects the reason wasn't as much suspected plagiarism as the user-friendly nature of the Lowell documents. For example, the Lowell researchers included finding charts for every single identified object. By the mid-1960s, the Giclas-Burnham-Thomas trio was publishing results from fewer plates at a time because "several observers using the results of this survey have indicated that they would like to receive our data more often," they wrote. Therefore, "we shall adopt a policy to publish more frequently beginning tentatively with every 10 plates completed." And that's exactly what they did through the ninth catalog. During the catalogs' heyday, more than four hundred published scientific papers cited the Lowell Proper

Motion Survey, according to the Smithsonian/NASA Astrophysics Data System, an exhaustive bibliography on astronomical research. The Lowell program ended up getting relied upon more than any other for subsequent work. For example, astronomers at Southern California's Mount Wilson Observatory used the Lowell Proper Motion catalogs for a long time to identify white dwarfs to inspect in their spectral studies. These days, the U.S. Naval Observatory is the chief organization still pursuing astrometry, that part of astronomy that measures the proper motion of stars as a function of time.

The three researchers completed the entire northern hemisphere in 1971 and published a summary. They also issued lists of positions for white dwarf star suspects that they'd found while hunting for proper motions. Some time after the Hyades paper, they made a regular habit of looking for proper motions a fifth of an arc-second or more. By the early 1970s they were finding so much on their plates that they issued one paper that covered what they had seen on only two. In 1978, they finished scanning the southern hemisphere, having found a total of 2,758 close stars there. And by 1979, their catalogs were in such high demand that they had run out of copies of results from the first one hundred plates. So they made more copies of the finding charts that they could send to researchers wishing to use them. The Lowell team had mapped nearby stars in the entire northern hemisphere and the southern hemisphere, too: nine thousand high-motion stars; two thousand asteroids; fifteen hundred white dwarfs; and a number of so-called variable stars, which change in brightness at regular intervals. Except for a tiny portion near the South Pole, they had wrapped up the entire sky. The National Science Foundation stopped funding the project, and it came to a close.

And thus began the sad end of Robert Burnham.

Even among the astronomers who worked with Burnham, few knew him. Wes Lockwood, who overlapped with Burnham by a few years, said the reason was simple: "He never talked to anyone." But Burnham did open up to a few people, and they came away feeling as if they had met someone very special. Nat White is another Lowell astronomer whose tenure overlapped with Burnham's. He remembers him as a kind and talented, if shy, man. Burnham lived in a do-it-yourself cabin from Sears and Roebuck on Mars Hill—the same one that Brian Skiff, an observer for several of Lowell's programs, occupies now. White lived nearby, and he recalls "very mystical music coming from there. He had an organ and he would play music just from his head." White says that Burnham also painted under the fluorescent lights inside his little cabin. "When you'd take them into the sunlight the paintings would have this mystical quality," he said. "I'd go over there once in a while. He loved kids. He felt that he could talk to kids better than adults, I think. If he would allow himself to talk to you, he would say amazing things. I don't think he was crazy. He was a recluse. He was uncomfortable in social settings." Burnham had found another friend in Thomas. During those long hours blinking plates, the two had many conversations, and Thomas never tired of listening to Burnham. "He knew so many things," he says. "He knew about poetry, literature, ancient history." He had Roman coins, coins from Judea, Christian coins." The coworkers teamed up to right things they found mishandled at the observatory. Some of the older astronomers, for example, didn't see the need to buy blinds for shielding windows in the rotunda when they showed slides for public presentations. "Bob Burnham went and bought drapes himself," Thomas said. "It may have rankled Henry, but we were desperate."

Even as he'd competed with Thomas to scour plates for the proper motion survey and discovered a handful of comets, Burnham had been pursuing another enormous project: the handbook of the heavens that he had started to envision shortly after a brief stint in the U.S. Air Force. By the time he was hired onto the proper motion survey, Burnham had filled half a dozen notebooks with his notes and drawings of the heavens. And by the time he self-published his *Celestial Handbook* in 1966, it was two thousand pages long. Eventually, the Dover publishing group took it on and sold it in three-volume sets. It was the only book of its kind when it was first released, peaked in popularity nearly twenty years later, and remains unparalleled as a night sky reference. Still, when the proper motion project came to a close, Giclas says, he didn't consider Burnham to be qualified to stay on at the observatory. He didn't have the education. Thomas had completed a geology degree at Northern Arizona University during the course of the project, so the astronomers put him to work on a variety of projects, including Ted Bowell's early search for new asteroids. Eventually, Thomas even went on observation expeditions to Hawaii and other places. But by the end of 1979, Burnham was out of a job.

At first, things didn't look so bad. Burnham was reportedly miffed about his termination from the observatory, but he was pulling in thousands of dollars in royalties from the book. He moved out of the lonely cabin on Mars Hill where he had been living a hermit's life for more than twenty years, and he rented an apartment in town. And then Robert Burnham disappeared. A thirteen-page, 1997 article in the Phoenix *New Times*, a weekly culture and entertainment newspaper, contains the most comprehensive biography of Burnham ever published. Its author, Tony Ortega, proposed that one reason Burnham was less than famous for his *Handbook* was a matter of identity confusion: around the time the book was popular, there was another Robert Burnham who edited *Astronomy* magazine, and many people assumed they were one and the same.

Once he moved off the Hill, he withdrew more and more—and did not seem to have the motivation or the wherewithal to establish new connections or pursue new opportunities. The royalty checks dwindled, and a science fiction novel he had envisioned never sold. At one point Burnham's truck was found off the road near Woody Mountain Road, south of Flagstaff. Thomas thinks Burnham may have suffered a stroke or an aneurysm while hunting for treasure in that area. That's the only way Thomas can explain what happened. Oddly, Burnham himself was later found wandering the beaches of California in poor physical shape. Around that same time, Thomas remembers walking into Burnham's apartment one day to find his friend's things scattered about and the bedroom window thrown wide open. "You've got to get up here," he remembers telling Burnham's sister Lydia on the phone. And that's the point in telling the story where Thomas has to stop because he is weeping so hard he cannot speak. "Poor Bob," he manages, before he recovers and changes the subject.

For his own reasons, Giclas does not even go that far. "Last I heard of him he was wandering the streets of Los Angeles and selling paintings of cats," Giclas says. He was referring to the time in the early 1990s—shortly before Burnham's death—when Thomas found him making likenesses of cats in San Diego's Balboa Park. Asked if Burnham was crazy, Giclas ignores the question without batting an eye: "Thomas was a grad student at Berkeley," he says. "He stayed on."

Chapter 5

TO WATCH THEM DANCE

Otto Franz is well into his seventies. He and his wife Gallina have been married for forty-four years. They have fun pointing out that he's from Austria, she's Russian, and their dog is a Weimaraner, a breed originally from Germany. Otto Franz exercises daily and works part time on his research at Lowell Observatory. There, he is winding down a research career that has spanned more than sixty years—a fitting time to witness a giant leap in his field that literally brings him to tears. He cried tears of joy the day before he flew to Prague in August 2006, to witness the presentation of present binary star data he had helped glean from the Hubble Space Telescope. The leader of his team, Todd Henry of the Harvard-Smithsonian Center for Astrophysics in Cambridge, would be giving the talk at a semi-annual meeting of the International Astronomical Union. Among other roles, the IAU has been the official naming body for astronomy since 1919. It was the very same meeting that would rock the worlds of astronomers and school kids alike when the group ousted Pluto from the lineup of true planets—but Franz wasn't focused on Pluto. In preparation, he had constructed a graph showing the luminosities and masses of ten faint binary stars. It was the first time a so-called mass-luminosity curve had been determined for such stars, and astronomers had been dying to see such a thing since they could see stars. Otto Franz is a grown man, a rational scientist. But looking at those data made him cry.

"As my career winds down," he explained, "I finally see this result that I've been working toward for a lifetime."

For most of Franz's youth, Austria struggled with occupation by the German Reich and its aftermath. He was in high school in the mid-1940s when "school disappeared behind the iron curtain," he said. "When it started again in 1945 after a hiatus of eight months, there were eighty in the class in one basement room. It was dismal, awful. Everything was awful—but we had excellent teachers." Franz said all his instructors enjoyed the title of professor even at the high school level. They were motivating and stimulating, he recalls. They'd all completed coursework for doctoral degrees and most had written dissertations. Beyond excellence in the classroom, Franz pursued two separate interests on his own: ornithology and astronomy. In his junior year of high school he had to make a trip to Vienna for some medical attention, and he decided to use his time there to visit university-level instructors in both fields. He asked for the time of a biology professor, whose main interest was ornithology, and a professor at the Vienna Observatory. "I never heard from the biology professor," he said. "But I got word back from the observatory that they'd be delighted to have me." One of the senior astronomers spent hours showing him around and answering his questions—and the cordial treatment continued through the time, years later, when he completed his doctor of philosophy degree and became their junior colleague. Starting then and continuing through the present, Franz has been focused on a single feature of the heavens: binary stars.

"Throughout my professional lifetime," Franz says of his work, "I have in fifty years used almost every technique that has ever existed."

Stargazers have always noticed that an uncanny number of stars seem to be close together, but until they had the tools to look closely, most people believed that was mere coincidence; they figured the stars were actually quite far apart and just happened to be in the same line of sight. In 1767, an English clergyman named John Mitchell started musing that there seemed to be an awful lot of these chance alignments. Almost fifty years later, William Herschel discovered the first known binary star: Castor, in the constellation Gemini. Within twenty years he had compiled the first catalog of 848 paired stars. Fast forward to the 1990s, when most researchers believed that about half of all stars existed in pairs. And in 2006, Franz says most stars occur in such associations.

Franz likes to say he studies binary stars partly for the same reason a mountaineer climbs mountains: because they're there. "It's a no-brainer," he says in his thick Austrian accent. "When most stars exist as binary stars we'd better study binary stars or else we're not studying stars. If we study star formation without looking at binaries…we're already not asking the right questions." Studying binary stars gives insight into the fundamental link between star mass and luminosity, which increases multiple times for every doubling of mass. The only way astronomers know even that much—and the only way they learn more about those parameters—is by studying paired stars. That is because measuring paired stars' gravitational effects on each other leads directly to an understanding of their mass. Lone stars give little indication of how massive they are. The quest to study stars in general sheds light on the formation of our own Sun, and it contributes to the ever-popular search for other stars that contain planets with the potential to harbor life. So interest in paired stars, as windows into these realities, has grown over time. And for nearly half a century, Franz has occupied a comfortable perch atop Mars Hill while he has surfed the gripping evolution of binary star research.

When he started his doctoral degree, Franz used a technique that had been in operation since the time of Wilhelm von Struve, a German-born Russian binary star pioneer who worked in the nineteenth century. The tool, a filar micrometer, is a crosshair mounted on a telescope's frame. Franz is quick to point out that he did not actually use the instrument for binary star research; his dissertation project was theoretical. He practiced using it to find positions of comets. Mainly, binary star researchers used the micrometer to get two characteristics of any binary pair: the separation of the two stars and their angle with respect to the celestial equator, which is the plane of the Earth's equator stretched out into space. That classic device is still rarely used by amateur astronomers, but binary researchers have moved on. Franz has progressed through a succession of increasingly impressive tools for spying on the intricate interactions of paired stars including ever-bigger Earthbound telescopes and, most recently, the Hubble Space Telescope (HST). The HST has been used to peer at binary stars so distant from each other that their mutual gravitational tug is almost unfathomable and, on the other end of the scale, stars so enmeshed that they "practically roll over each other's surface," Franz says. It has uncovered drama of celestial proportions. Some coupled stars steal each other's mass, while others abandon each other in explosive death displays.

While he was at the Vienna Observatory, Franz made observations as an assistant to his major professor and the observatory director, Josef Hopmann, on a 26-inch refracting telescope. The telescope was the biggest in the world for a few years after it was built, half a dozen decades before Franz peered through it. And he used a desktop hand crank calculator to estimate the mass, intrinsic brightness and temperature for four hundred binary stars out of forty thousand that had been identified.

That is to say, with respect to binary stars, astronomers had not come all that far from where William Herschel put them in 1802. Certainly, their main tool—the filar micrometer—hadn't evolved. When it is used for observing binaries, a telescope fitted with a filar micrometer works in two ways to tell astronomers what they like to know. It starts with a fixed crosshair mounted at the telescope's focus. The frame also has guide rails that support a slide, which carries a second, parallel crosshair back and forth across the line of sight. To measure the distance between two stars, an observer twists the whole assembly so each of the crosshairs bisects a different member of the pair. The distance between the crosshairs is measured to give a reading of the distance between the stars, in units called arc-seconds. The micrometer is also fitted with a rotating outer circle, divided into degrees, that captures the east-west and north-south positions of each star.

"For two hundred years, that's how they did it," Franz says. "This was the workhorse of binary star research."

The micrometer only helps characterize a star pair, of course, if it is indeed a pair. Paired stars always show the influence of their partners. Visual binaries can actually be seen orbiting their shared center of mass. Eclipsing binaries dim at regular intervals because their partners pass between them and observers on Earth. The A star in Orion's Trapezium is one famous eclipsing binary. Sometimes only one star is visible, but it wobbles because of the gravitational tug of its unseen partner. Those are called astrometric binaries.

The filar micrometer is only useful for a pair of stars where both stars are clearly visible to researchers on Earth. Sometimes the companions are too small to shine, but dance with the visible star just the same. One of the reasons astronomers became aware that there are so many paired stars, even before telescopes were powerful enough to see some of their partners, is that they have developed the ability to detect stars' motions. They use simple geometry and a parameter called parallax: the apparent shift of an object as seen by an observer, due to the motion of the observer. Franz happily demonstrated this in his office one day, using both eyes and his own thumb at the end of his outstretched arm. Closing first one eye and then the other, he noted that his thumb appeared to move back and forth even though he was holding it perfectly still. "If you measure the displacement angle and you know the distance between your two eyes, simple geometry tells you how long your arm is," he said. When observers want to know how far away a star is, they can simply take measurements of its position two seasons apart—summer and winter, for example—when the Earth is at the opposite ends of its orbit around the Sun. The scale is much larger, but the math is just as clean. Unless, of course, there's an unseen companion buggering things up.

When astronomers measure stellar parallaxes, they often do it for several years and their measurements, if connected with a line, would make a regular, back-and-forth pattern—just as a thumb held out from your face would go back and forth, back and forth, by the same amount each time you switched eyes. But sometimes, there is an extra wiggle in the curve. It's called perturbation, and it shows astronomers there is some other object tugging on the orbit of the star. Just like visible paired stars, astronomers have been seeing the phenomenon of perturbation for well over a century. In 1846, Neptune was discovered from perturbations in the orbit of Uranus. Around the same time, Friedrich Wilhelm Bessel announced perturbation in the proper motions of Sirius and Procyon, and chalked up both perturbations to the presence of invisible companions. Both were spotted visually when big enough telescopes came along—the companion of Sirius by Alvan G. Clark in 1862, and Procyon by J.M. Schaeberle in 1896. German astronomer Johann Gottfried Galle used perturbation in the orbit of Uranus to discover Neptune in 1846. And although Percival Lowell's sought-after Planet X was never really found—Pluto was too small to fit the bill—he had Tombaugh searching for it based on perturbation he observed in the orbits of Uranus and Neptune.

For years, all the results were computed with desk calculators and logarithmic tables. Observations took place over decades and even centuries, yielding huge systems of equations that would reveal the extra orbital wiggles of paired stars with hidden partners. "Nobody has seen it," Franz says of such unseen companions, "but it's got to be there because gravity doesn't lie." Now, of course, computer programs do all the work. He says: "Easy. Now you put these numbers in and seconds later you have your answer and a plot." Telescopes have also evolved to handle binary star work.

Early parallax studies were done mostly by refractors, which use lenses to focus light from celestial objects. It took reflectors, which focus the light using mirrors, a while to rise to the task. One of the first reflecting telescopes suited to studying stellar parallax, and dedicated for that purpose when it saw first light in 1964, was the 61-inch reflecting telescope at the United States Naval Observatory in Flagstaff. Franz eventually got to use that telescope.

• • • • • • • • ◦

Franz finished his degree at Vienna University in 1955 and left within days for Northwestern University's Dearborn Observatory, just north of Chicago in Evanston, Illinois. He'd met his boss, Kaj Strand, at an IAU meeting in Dublin, and Strand had hired him on the spot, he remembers. Franz stayed in Illinois for only three years before following Strand to the U.S. Naval Observatory in Washington. Once there, Strand became too busy with administrative work and the planning of the 61-inch telescope in Flagstaff to work much on his program to observe binary stars. That included his semi-annual binary star observing runs, using the 24-inch refractor at Lowell Observatory. So Franz's first visits to Flagstaff, in 1959, were at the behest of his boss. Franz remembers he drove "one of the first Volvos in the Chicago area," and took it to Washington when he moved there. It was the same car he drove back and forth to Flagstaff twice a year for the observing runs, because he had to haul delicate equipment across the country each time. He recalls the first trip in February of 1959, when he used Old Route 66 in the middle of a huge snowstorm.

"It was snowing like nobody's business," he said. "I didn't know where the town started." So he stopped at the first lodging he saw, which ended up being the Snowbowl Motel—directly across the street from the tiny railroad depot and near the heart of downtown. The motel wasn't much then—a single line of rooms at the base of a small cliff—and it has since been torn down. But it offered a reasonable weekly rate, and for those first few visits, Franz called it home. Eventually he would stay at other motels, or in guest quarters at Lowell Observatory itself. "For heaven's sakes," he declared during one of our interviews more than half a century later, genuinely startled, "it was right here. God, that never occurred to me. This was a bedroom, and now it's my office." Franz kept up the semi-annual commute during the years 1959 to 1965. Meanwhile, he and Gallina married in 1962, and she soon gave birth to the first of their three children, a son named Nick. Franz said he liked Flagstaff, but it never occurred to him that he would one day make it a home. In fact, he had become acting director of what was then called the Astrometry and Astrophysics Division at USNO and it showed him he definitely didn't want to give up research for administrative tasks. So he was interviewing for positions at Yale University, Swarthmore College (the home of Sproul Observatory), and Van Vleck Observatory at Wesleyan University in Connecticut. But the universe had other plans. By then Strand's predecessor at the USNO, an astronomer named John Hall, had become director at Lowell Observatory. Hall had also completed his graduate studies at Yale and got wind through the grapevine of Franz's efforts. "When he heard I was planning to leave the Naval Observatory, he called me in Washington and said, 'Please don't make any decisions until I've had a chance to talk to you,'" Franz recalled. Hall had hired Franz once already, because Franz and Strand had come as a package deal when Hall was interviewing his replacement at the USNO. He often joked that he hired Franz twice: the second time was at Lowell. By then the 61-inch telescope was operational at USNO's Flagstaff station. Kaj Strand had originally conceived of the telescope, and his name is the one linked to the telescope's best known early finds. It was named after him in 1997.

· · · · · · · · ○

Starting in the 1970s with the work of French astronomer Antoine Labeyrie, astronomers began seeing binary stars more clearly with a technique called speckle imaging. It works by taking a series of quick, short-exposure images like a movie. In each frame, the blurring effects of the atmosphere spread the light out over a fuzzy patch, but within that patch there are bright points, called speckles. Franz used speckle imaging with colleagues from Georgia State University throughout the 1980s, and with other colleagues since then. Speckle imaging is quick. In a 2006 paper, Elliot Horch at the Rochester Institute of Technology, along with Franz and several co-authors, mentioned in a report on speckle observations of binary stars from the southern hemisphere that "we typically acquired 1000 or more speckle frames of data on a target, which took approximately two minutes, including readout time." From frame to frame, the position and intensity of the speckles changes, and studying the speckles in each short exposure allows researchers to piece together a high-resolution image. It helps resolve very close binary stars that would otherwise be blurred together in images taken from the ground. And it allows astronomers to determine the masses of such systems and compare them with theoretical predictions. In recent years, Horch has been taking speckle observations, mainly of star pairs, at the WIYN 3.5-m Telescope at Kitt Peak National Observatory near Tucson. He has been collaborating with Bill van Altena at Yale University—the very same person who sparked Luyten's ALL CAPS wrath in the 1960s—and Zoran Ninkov, also at RIT. Horch and Franz have made several trips to the Lowell-Tololo telescope at the Cerro Tololo Inter-American Observatory, where the pair measured binary systems in the southern hemisphere. Some of the targets for their observations were identified by the European Space Agency's Hipparcos astrometric satellite, which operated in space from 1989 until Earth-bound astronomers stopped communicating with it in 1993. The 1.4-ton craft pegged the locations of one hundred thousand stars, including thirty-four hundred new double stars and thousands of other suspects.

· · · · · · · ·

The United States' own record of placing observing satellites in space began in earnest starting just after 1957, when the USSR launched Sputnik. "The American public was horrified by that," said David Klanderman, who constructed a 2005 exhibit about the ongoing international space race at Tucson's Pima Air and Space Museum. "But the official U.S. response was unbelievably calm: 'Let the other guys set the parameters. Then we'll do whatever we want.'"

The main advantage of space-born observatories is that they get past the interference posed by the Earth's atmosphere, including heat-related turbulence and the absorption of radiation at many wavelengths. Shortly after Sputnik, NASA sent up a series of Orbiting Astronomical Observatories. The first, launched in April of 1966, was OAO 1, meant to measure the use wavelength absorption and emission—a window into chemical composition—of stars, planets, nebulae, and the random matter between stars and planets. It spied on these objects using the regions of the spectrum from visible light to gamma rays. OAO 2 was launched nearly two years later. It carried eleven telescopes and performed X-ray, ultraviolet, and infrared observations of stars. In 1970, a satellite called OAO B failed to launch. And in 1972, OAO 3 focused on stars in the ultraviolet wavelengths. Mostly, members of the OAO series were learning experiences. But scientists were clamoring for more powerful satellites. By the late 1960s, momentum was gathering for a large orbiting space telescope. And by 1969 NASA gave a nod, although not a creative name, to the Large Space Telescope (LST) project. After Armstrong's "giant leap for mankind" on the moon in 1969, funding for NASA space programs began to dwindle, putting the LST program in jeopardy. LST planners had to design the telescope under budget constraints nearly from the word go, but they got just as much pressure from scientists to move forward. In an introduction to Peter van de Kamp's 1981 book *Stellar Paths*, Jean-Claude Pecker writes, "Peter van de Kamp argues vigorously for the maintenance of long focal-length refractors for the uninterrupted pursuit of their programs, and he is right, without a doubt. But, without trying to predict what astrometry at the end of the century will be, it is clear … that extraterrestrial instruments will be needed —astrometric satellites, loudly called for today by both astrometrists and astrophysicists, in harmonious concert." Astrometry is a type of astronomy that focuses on stars' distances and proper motion. In the end, the agency designed a series of four satellites, or Great Observatories, that would collectively cover the spectrum of observable wavelengths not easily detected on the ground. The first was the Hubble Space Telescope, launched in 1990. The Compton Gamma-Ray Observatory orbited from 1991 through June of 2000 and collected data on some of the most violent physical processes in the universe, characterized by their extremely high energies.

The Chandra X-Ray Observatory was deployed from a shuttle and boosted into a high-Earth orbit in July 1999 and continues to observe black holes, quasars, and high-temperature gases. The fourth and final Great Observatory is the Spitzer Space Telescope, launched in 2003. It makes observations in the thermal infrared of regions of star formation, the centers of galaxies, and newly forming planetary systems. Infrared can also focus on cooler objects in space—such as stars too dim to be detected by their visible light, extrasolar planets, and giant molecular clouds.

Of these, astronomers have gotten the most from observations by the Spitzer and especially the Hubble Space Telescope. The former has for the most part hummed along without drama, save the ever-present NASA funding rollercoaster, periodically sending back data that pleases the public and satiates astronomers. Hubble is a different story. Its decade-long preparation phase, and even more so the tumultuous time following its launch, was so volatile that one of the main players, Eric Chaisson, wrote a book afterwards called *The Hubble Wars* that reads like the painstaking account of someone's really bad dream. It is filled with horrid tales of scientists acting like grandiose children, malfunctions that cost NASA countless thousands of dollars and squandered years of time for astronomers, and a sometimes-irresponsible media frenzy that kept the whole circus going. NASA's funding pressures seem to have been only part of the trouble. Awkward points in the design and engineering came from that, from an embarrassingly shoddy quality control program, and an unwieldy management scheme. In constructing their observing programs, scientists had to navigate technologies that had been optimized not for research but for defense programs. "In truth," Chaisson writes, "a fair amount of Hubble's technological design and fabrication derive from Project

Keyhole, a 'dark-side' enterprise that has during the past two decades orbited a virtual fleet of telescopic intelligence gear." For all of these reasons, Hubble mutely orbited Earth for most of its first three months of deployment. It was three years before a servicing mission fixed the majority of problems on Hubble that interfered with its ability to make images of celestial objects that didn't have to be doctored. That left the public waiting a long time for pretty pictures from outer space, and it left observers like Otto Franz waiting some time for the astrometric measurements—including masses and luminosities of binary stars—they were so eager to see. Franz says the most infamous glitch, a flaw in the main mirror, never did get fixed. But astronomers have learned to live with it.

The HST is designed with a three-tiered system to help it latch onto its observational targets. The first is a set of gyroscopes that steer the vehicle with rough accuracy toward a certain swath of the sky. Then, three miniature telescopes hone it in a little more. They have fields of view ten times the diameter of the moon as viewed from Earth. Then comes the amazing power and accuracy of the Fine Guidance Sensors, which are meant to be able to lock in on one among millions of stars and to respond to any apparent motion of those stars with unparalleled accuracy—"about the width of a human hair at five miles," Chaisson writes. The intention of the telescope designers was that two out of the three Fine Guidance Sensors would be used at any one time for steadying Hubble's aim. The third would be available for making precise observations of stars, including binaries. But the guidance sensors suffered on multiple fronts. Their star maps were incomplete and choc full of errors, so they sometimes got distracted and locked onto the wrong targets. And an unrelated engineering glitch led to a perpetual wobble that made any sort of precise focus almost impossible. Finally, several months after launch, Hubble sent astronomers some relief when it peered neatly into R136, a large star cluster in the Large Magellanic Cloud. Shortly after that, it sent back the clearest images to that point showing the relationship between Pluto and Charon. Hubble also helped capture the once-in-a-lifetime white spot on Saturn, caused by intriguing turbulence in that planet's atmosphere.

Throughout this time, the HST has been orbiting three hundred miles from the Earth's surface. The span of the HST's investigations is vast, including essentially all areas of astronomy: cosmology, galaxies, intergalactic gas, stellar populations, interstellar medium, hot and cool stars, star formation, solar system, quasars, and active galactic nuclei. Every day, it sends ten to fifteen gigabytes of new data back to Earth. Still, scientists hungry for its observations retrieve the data five times as fast as it comes. The space telescope had been used for nearly fifty-four hundred research papers in peer-reviewed scientific journals by the summer of 2006, with six hundred fifty-one of them published in 2005—more than in any other year.

As the functioning Hubble Space Telescope evolved, so did the scientific partnerships that would use it. Franz and colleagues, including Laurence Fredrick from the University of Virginia and Yale's van Altena, started planning for the astrometric use of the Hubble Space Telescope as early as 1972. The three met in Green Bay, Wisconsin, talked van Altena into being the chairperson, and wrote a joint proposal. They wanted to use the Fine Guidance Sensors to measure distances to stars and help tease out many binary stars that still eluded description. But NASA wanted to use them to obtain a systematic catalog of reliable guide stars that would guide the telescope. A team led first by William Jefferys at the University of Texas—and later by his colleague, G. Fritz Benedict—was eager to direct the FSGs in providing that sort of service. Franz says Nancy Roman, a prominent astronomer and early NASA scientist, was able to reconcile the two proposals.

"She had the vision there was important astrometric science perhaps possible. She engineered a marriage between the two teams. Ultimately, what the Texas people proposed to do, NASA wanted to do in house. So they all shifted over into doing science."

So, starting in the mid-1990s, Franz was part of a team that began publishing papers about all sorts of binary stars. Hubble allowed them to spy on Wolf 424 B, the first known star pair to comprise two low-mass brown dwarfs. Brown dwarfs are hard to see because they are below the threshold at which stars are able to sustain themselves with hydrogen fusion, weighing in somewhere between stars and giant planets. The researchers, in a 1998 paper in the *Proceedings of the Astronomical Association of the Pacific*, touted the find as another step toward understanding "star formation and the faint end of the luminosity function ... at the low-mass luminosity boundary between stars and the domains of giant planets," adding that "large numbers of unobserved, low-luminosity brown dwarfs may provide some of the 'missing mass' inferred by both galactic dynamics and cosmology." That same year, a large team of authors including Franz and his Lowell Observatory colleague Lawrence Wasserman published in the *Astronomical Journal* a celebratory account of the very first binary star orbit nailed down with HST's fine guidance sensors: the M dwarf binary Wolf 1062. Wolf 1062 was on a list of suspected perturbation binaries Franz contributed to one of the very first astrometric research proposals for Hubble. The telescope found the companion almost immediately, and "set the stage for everything we could do afterwards," Franz says. Wolf

1062 was notable for a different reason: at the time, it numbered among the eleven lowest mass objects in space for which accurate masses had been determined.

But what Franz is most proud of—and what he was so excited to see presented at the IAU meeting—is a relationship that is just now coming clear between the mass and luminosity of stars. He reflects back to the early 1980s, when astronomers working diligently for a century and a half had determined accurate masses for only nine visual binary systems, with margins of error up to thirty percent. "That was the best information available in 1980, period, for visual binary stars," he said, jabbing the air with his finger for emphasis. Plus, for masses less than two tenths of a solar mass, there was no known luminosity at all. So astronomers were worlds away from having an established relationship that would allow them to look at a star of a given brightness and predict its mass, especially for faint stars. That is, they were lacking a standard mass-luminosity curve. And that is where the Hubble Space Telescope came in. The Harvard-Smithsonian's Todd Henry proposed to determine the mass-luminosity relationship at the end of the main sequence. That means if you put all known stars on a graph of brightness, Henry wanted to study the ones at the faint end of the scale. He teamed up with Franz, his longtime Lowell colleague Wasserman, and the University of Texas's Benedict. The team has been peering at a dozen binary star pairs using the space telescope's fine guidance sensors, which can hone in on them with orders of magnitude of more resolving power than current ground-based telescopes. Early in the effort, they nailed down orbits for all twelve pairs within a few percentage points of accuracy. And the *piece de resistance*, the graph that gets Franz misty-eyed, is a plot with a line superimposed on it—representing a mass-luminosity relationship for stars on the faint end of the main sequence. All twelve binary pairs are less than half the mass of the sun, and until now would have been impossible to analyze like the HST has. "We are finding a structure here that was not known before," he says, before emotion makes him pause. "It's a nice finale for a career ... but I am not completely done yet." Franz says despite the mass-luminosity relationship that has finally emerged from the team's data, there are outlying observations that will still need to be explained. But for the first time, there are now enough data to hand over to the theorists and physicists, who will best be able to interpret the fine points. And he, for one, is eager to watch the next lines of the stars' stories unfold.

Chapter 6

OUR SUN, OUR STAR

"It cannot too often be repeated that the Sun is the only star whose phenomena can be studied in detail." George Ellery Hale

For as long as we have had the ability to chronicle our thoughts, human beings have harbored fantasies about life beyond our own realm. The Greeks made up myths about the creatures living outside their shadowy borders. Entire subcultures have sprung up around UFOs and movies about all manner of extraterrestrial beings. We hold a collective, secret conviction that *Star Trek* is a plausible scenario ahead of its time.

Through the years, science has continually tried to make good on the public's fantastic obsession. The efforts vary in credibility. Percival Lowell's insistence that he saw canals on Mars, that they had been created by intelligent life, and that he could see seasonal fluctuations in the vegetation around them, proved misguided.

The Search for Extra-Terrestrial Intelligence, or SETI, got its start in 1959 when Cornell physicists Giuseppi Cocconi and Philip Morrison published an article in *Nature* touting the potential for using microwave radio to communicate between stars. Meanwhile, a young radio astronomer named Frank Drake, without knowing it, was planting the seeds for the 1990s movie *Contact* that would feature Jodie Foster in the role of a scientist obsessed with extra-terrestrial life. In the spring of 1960, Drake was aiming an eighty-five-foot antenna in West Virginia toward two nearby Sun-like stars, for the world's first microwave radio search for signals. His single-channel receiver was tuned to the same frequency preferred by Cocconi and Morrison: the twenty-one cm (1,420 MHz) line of neutral hydrogen. Drake didn't detect any other-wordly signals, but his so-called "Project Ozma" sparked the interest of others in the astronomical community, starting with the Russians.

Efforts to detect life within our own solar system, while still very active, have uncovered little over the years to get excited about. The Mars Phoenix Lander, scheduled for launch in August 2007, is the first in NASA's Scout Program, comprising a series of probes designed to be low-cost complements to major missions in the agency's Mars Exploration Program. Phoenix is headed by Peter Smith of the University of Arizona's Lunar and Planetary Laboratory. It is designed to measure volatiles, especially water, and complex organic molecules in the arctic plains of Mars. Organic molecules contain carbon, the stuff of life. George Bell of Arizona State University wrote an academic tutorial in 1997, "The Search for the Extrasolar Planets: A Brief History of the Search, the Findings and the Future Implications," in which he pointed out that Jupiter's moon Europa could have a liquid ocean under the surface and could harbor living things. NASA also favors Jupiter's moon Callisto as possibly hosting an ocean, and therefore life. But for the most part, the search has moved away from our Sun to other stars. And while our technology is not yet good enough to actually see them, astronomers have used a variety of methods to indirectly observe several hundred planets orbiting stars besides our own.

Two century-old frameworks are still used to talk about stars. The first is Annie Jump Cannon's system of classifying stars by their spectra. Her O, B, A, F, G, K, M system of spectral classes, ranging from hottest to coolest, was completed in 1911 and adopted by the International Solar Union in 1913. It is still used, and still often remembered by the famous mnemonic device "Oh, be a fine girl—kiss me." Our own Sun lies in the G range, with a surface temperature of approximately 5,500 Kelvin, giving it a white color that appears yellow to us only because of the scattering effects of our atmosphere. The second framework is the classic Hertsprung-Russell diagram, conceived independently by Henry Norris Russell and Ejnar Hertzsprung and published in 1913. It is basically a graph that plots Cannon's spectral classes against the added stellar properties of absolute magnitude, luminosity, and surface temperature, all of which decrease with a star's age. Again, our own Sun falls roughly in the middle.

A variety of Earth- and space-bound techniques are employed to detect planets around other stars. The Earth-bound methods are intuitive enough. The so-called transit method observes the shadows of planets as they pass between Earth and their host stars. Microlensing is a bit more complicated, honing in on the distortion of a star's light caused by a foreground star. The light ends up distorted into an arc, bent around the intervening or lensing star by the time it reaches viewers on Earth. If the lens star has a planet, that planet can also act as a lens, producing a brief secondary arc lasting days for a Jupiter-size planet or hours for an Earth-size one. The Probing Lensing Anomalies NETwork, or PLANET, is a consortium of five telescopes distributed around the Southern Hemisphere and dedicated to searching for lensing events. Microlensing is the only known ground-based way to find other planets as small as Earth. Finally, astronomers can intuit the presence of Jupiter-size planets using the Doppler wobble, the gravitational tug of the unseen planets on their host stars. The wobble technique is the oldest method to yield results, dating back to a controversial little star discovered by E. E. Barnard in 1916. All Barnard noticed was that it had an unusually large proper motion. Peter van de Kamp, the same person who introduced his 1981 book *Stellar Paths* with a call for space-based observatories, carried Barnard's Star farther than that. He spent much of his career documenting— and defending—his detection of a wobble in Barnard's Star and his belief that two planets were causing it. Numerous astronomers, including Robert Harrington at the USNO's Flagstaff observatory, tried to confirm van de Kamp's sighting of the wobble and could not. Others did. As of van de Kamp's death in 1995, the jury was still out, and his life's work still has not been confirmed.

The Doppler wobble decreases—and becomes harder to see— when it is caused by small planets that orbit farther away. That is the main reason that about forty percent of known exoplanets are Jupiter-size giants in close orbit of their stars. Avi Mandell, an astrophysicist at Penn State and a researcher with the NASA Goddard Space Flight Center in Greenbelt, Maryland, says that sample is biased because detection of smaller planets is so much more difficult. As space scientists build ever bigger and more powerful telescopes, Mandell thinks higher percentages of small and distant planets will be found.

• • • • • • • ◦

Brian Skiff, fifty-one going on thirty-five, emerges from his build-it-yourself cabin on Mars Hill around ten or eleven each morning. His straight, light brown hair hangs around his ears and he walks with quick, short strides—the same way he speaks. Clad in a T-shirt that's always tucked into loose-fitting, casual pants, he walks the twenty yards across the parking lot to the Slipher building that contains the small library and some of the astronomers' offices, including the one he shares with fellow researcher Bruce Koehn. He turns on the computer before he gets in his little white Corolla to make the trip down to Macy's Coffeehouse to grab one pastry for breakfast and another for the afternoon. By day he sits in front of a computer, painstakingly checking and revising old star positions for a comprehensive catalog effort called SIMBAD, which is run by the Centre de Données astronomiques de Strasbourg, in France. But that is just what he does by day, or when it is cloudy. Nobody bats an eye when he stumbles into work so late because on clear nights he is up until first light, making observations in a telescope dome atop Anderson Mesa. Skiff lives a life not many others, even astronomers, would fight for. His cabin is the very same one vacated by Robert Burnham before that man's sad end. The cabin suits Skiff, even though its history gives him the "willies" sometimes. The harder part is the observing. For twenty years, he has endured strings of long, isolated nights in the telescope domes atop Anderson Mesa.

"It is unbelievably boring and physically demanding," said Wes Lockwood, one of the astronomers for whom Skiff takes data. "It's cold up there."

Skiff has made observations for two main projects: Ted Bowell's ongoing work to detect asteroids with orbits that could sling them into Earth, and Wes Lockwood's long-term studies of our own star.

Lowell astronomers first started getting curious about the Sun's variability around 1950, and Henry Giclas took a few measurements then. At the time, solar research was largely in the purview of Mount Wilson Observatory. George Ellery Hale and his staff started the observatory's solar program in 1904, and within a few years Mount Wilson was a world leader in astrophysical research. The group was the first to analyze spectra of sunspots, proving that the dark areas are cooler than the rest of the solar surface. This and other information would eventually lead to the understood relationship between the luminosity and spectral classes of other stars. As a student, Hale invented a spectroheliograph—a modified spectrograph for photographing the Sun in the light of a single element—and from Mount Wilson it took daily measurements of the spectral signatures of calcium and hydrogen on the Sun. Astronomers there also chronicled the Sun's rotation and its Doppler shift, and Hale uncovered early evidence of magnetic activity on the Sun, for the first time showing the Earth was not the only celestial body subject to magnetic forces. Eventually, he showed that the solar cycle was a twenty-two-year magnetic cycle.

Mount Wilson was also the birthplace of long-term investigations into the Sun's overall variability. When the observatory was first established, Hale invited the Smithsonian Institution to send out an expedition to measure the amount of radiation received from the Sun in the clear, dry mountain air, and to look for variations. That type of work went on for decades, employing bigger and bigger telescopes. The first two, the Snow telescope whose arrival from Yerkes Observatory coincided with the observatory's founding, and the 60-foot tower telescope completed two years later, are mostly used for education now. Only the largest, the 1912, 150-foot tower telescope, is still used by the Division of Astronomy and Astrophysics of the University of California in Los Angeles to investigate long-term changes in solar magnetic activity and short-term oscillations in the Sun. The bulk of long-term solar investigations wound down at Mount Wilson in the mid-1960s because it seemed to be a dead end. But weather and climate scientists really wanted observers to start it up again. So in 1970, the then-director of Lowell Observatory, John Hall, wrote a proposal to the National Science Foundation for a five-year program. He hired Mike Jerzykiewicz, and when Lockwood took a job at Lowell, they overlapped for several years.

"He taught me everything I know, then went back to Poland, and he disappeared behind the iron curtain," Lockwood said of Jerzykiewicz. "He and I had very little contact until the fall of the Soviet Union." By the time he arrived at Lowell, Lockwood had completed a doctoral dissertation in infrared astronomy at the University of Virginia. He had spent the interim working at Kitt Peak National Observatory, fifty-five miles southwest of Tucson. When he started at Lowell on the solar variability project, he said, "I thought it was a waste of time, and when it's over, I'll do something else. Thirty-three years later, I'm still doing it." Lockwood's work exemplifies a type of long-term science that is almost unheard of in the larger astronomical community, but that distinguishes Lowell.

Right away, Lockwood and Skiff started taking brightness measurements. They were not seeing much from direct looks at the Sun, and it was not until a few years later, in 1980, that the Solar Maximum Mission made measurements from space revealing that there is just a tenth of a percent short-term change in the Sun's brightness. "That's such a small amount it's impossible to see from Earth," Lockwood said. That craft launched on Valentine's Day in 1980 and carried several scientific instruments, which provided new insights into the nature of solar flares. SolarMax was rescued and repaired by a 1984 Space Shuttle *Challenger* mission—the first time any spacecraft was repaired and turned loose in space—and reentered the Earth's atmosphere in December of 1989.

Realizing that Earth-based observations couldn't reveal much about the Sun's brightness, Lockwood and his colleagues took an indirect approach: they started studying the variability of our own Sun using other planets. That work became a purely planetary science program and has been supported in recent years by NASA.

In the mid-1980s, Lockwood secured a contract with the U.S. Air Force to investigate the Sun from yet another angle: how much change was apparent on stars like the Sun. Lockwood and Richard Radick of the National Solar Observatory in Sunspot, New Mexico, with Skiff as the observer, worked on that project for fifteen years and learned that our Sun may be unusual in being less variable than its peers in the heavens. As of that time, Earthbound measurements could only capture variation three times that of the Sun and higher. No astronomer is about to criticize the Sun; after all, it allows us all to live. But Lockwood, for one, would have enjoyed better luck in that work if it would shake things up a bit. That project ended at Lowell in 2000, but it continues using automated telescopes at Fairborn Observatory in Sonoita, Arizona. It is now led by Greg Henry of Tennessee State University.

In addition to monitoring its overall brightness, Lockwood and Lowell astronomer Jeff Hall have been using spectroscopy and the 42-inch telescope at Anderson Mesa to track magnetism of the Sun and other stars. The older stars are, the less active they are. For stars a few million years of age or younger, variability is a few percent and very easy to measure, Lockwood says. They have been watching twenty-one hundred stars—and of those, the Sun appears to be middle-aged based on its magnetic activity. There are a handful of close matches. By studying the whole lot of them, Lockwood and his colleagues would like to be able to estimate how much the sun could have varied in the past and make a projection for the future. The bottom line from that work is our Sun seems to hardly change at all—even compared to stars similar to it.

Our Sun also stands out among other stars because it emits a comparatively modest amount of ultraviolet light. More might mean that everything in its path—including Earth—would stay fried to the point where life might not exist. "Our star happens to be very nice to us," Lockwood said, "but there was this funny thing with the Little Ice Age."

The Sun normally shows signs of variability, including its twenty-two-year magnetic cycle and its eleven-year sunspot cycle. It ranges from minimum to maximum activity, represented by a peak in sunspots and flares. The Little Ice Age was a period of cooling occuring after a warmer era known as the Medieval climate optimum, a period of unusual warmth in Europe between the tenth and fourteenth centuries. There is disagreement about the timeframe of the Little Ice Age, but it likely started around the fourteenth century and lasted several hundred years. Most climatologists agree that there were three unusually cold periods lasting about a century each and separated by brief interludes of warmer weather. Most believe the effects were limited to Europe, but there is some evidence they reached into other parts of the globe, including North America and even parts of the southern hemisphere. During this time, according to NASA publications, access to Greenland was largely cut off by ice, and canals in Holland routinely froze solid. Glaciers advanced in the Alps, and sea ice increased so much that no open water flowed around Iceland in the year 1695. Those effects occurred at the height of one of the cold periods, lasting from 1645 to 1715. Called the Maunder Minimum, it is believed that all those dramatic events coincided with a decrease in the total energy output of the Sun. For evidence, supporters of that theory point to Galileo Galilei and the world's first telescope.

The telescope was invented in 1609. By 1611, Galileo was making drawings of lower sunspot activity leading up to the Minimum. He—and the astronomers who later confirmed his findings—saw only about fifty sunspots over a thirty-year period. Normally, they would have seen closer to fifty thousand. "You can't ignore that," Lockwood says. And he thinks Galileo's observations reach far beyond the Little Ice Age: "If we knew how often the sun was going to have these episodes where there's very little activity, we might be able to predict the climate." That's the most important lesson from the Little Ice Age: that the Sun has the potential to influence the Earth's climate. For the people who study climate change, the type of long-term data Lockwood has collected is useful, along with growing records of the Earth's temperature and the abundance of carbon dioxide. So far, Lockwood says, the researchers who have the most credibility and are the most politically neutral say the Sun will contribute twenty-five percent of the Earth's climatic variability on a one hundred-year timescale. "Back before men started screwing up the atmosphere, it might have been bigger, up to fifty percent," he said.

Neither Lockwood nor his colleagues claim to be experts on the Earth's future in terms of climate change. But their patient work is yielding the puzzle piece that's likely to be contributed by the Sun. And within that piece fits the work to chronicle changes in brightness on other planets. Like the rest of Lockwood's studies

Chapter 7

SEEK AND DESTROY

Deep Impact was a newspaper reporter's dream. It was a blessed opportunity to command readers' attention with imagery that can only be described as primitively erotic—to craft playful sentences around terms like "penetrate" and "slam into" without getting canned. Not batting an eye, editors at Tucson's major daily newspaper, the *Arizona Daily Star*, ran a pre-impact story that led with the sentence "Everybody wants a piece of this action."

Perhaps the astronomical community didn't have its mind at all in the gutter when it made plans to send the washing-machine-size probe on a two hundred sixty-eight million-mile crash course toward the comet Tempel 1. NASA's Deep Impact craft was launched from the Kennedy Space Center on January 12, 2005 and got within shooting distance of Tempel 1 by July 2. That is when the copper-tipped, eight hundred twenty-pound probe separated from its mother ship and made a beeline for the comet, which astronomers likened to a four-mile-wide potato in space.

"There's a little bit of something destructive in everybody, where you want to hit something. That's satisfying about this mission," said a jubilant Tony Farnham, a co-investigator on the Deep Impact Mission from the University of Maryland. Farnham observed the impact from the 4-meter telescope at Kitt Peak near Tucson and gave a public talk to a sold-out crowd of about fifty people who drove up the mountain to watch the impact surrounded by one of the country's highest concentrations of powerful, ground-based telescopes.

The visitors, who had paid forty-five dollars apiece for the evening, started scanning the skies with binoculars and telescopes just after dark on that warm night—July 3, 2005. Meanwhile, more than two hundred fifty professional scopes, including four at Kitt Peak and three at Lowell Observatory, were also aimed at the faraway blast. Inside the visitors' center during the hours leading up to impact, astronomers on T.V. used materials like talcum powder to illustrate various ways the unprecedented collision might look. As it turns out, the people who chose comfortable seats indoors got a better view of the real thing than the ones standing outside with binoculars. The visitors had a better vantage point even than a handful of scientists nearby in the 4-meter telescope, who abandoned their computers and spectra at the moment of impact to dart outside and scan the sky in vain hope of seeing a flash.

The impact was an immediate success, producing a flare that brightened Tempel I by six times. Astronomers roared with glee when it hit, and the light show rendered NASA scientists in the control room speechless before they were able to utter words like "spectacular" over a national teleconference line. But it was not bright enough to be visible from Earth with the naked eye. And the plume of cometary dust spewed so violently that it obscured the very crater astronomers created and had hoped to investigate.

The entire mission cost more than three hundred million dollars, and astronomers say it was an important experiment that took advantage of a rare window into the outer reaches of the solar system, where material is kept pristine because of deep-freeze conditions far away from the Sun. Such objects can give insights into the condition of the solar system at its origin, they say. They also say lessons from a project like Deep Impact could help spare the human race from certain catastrophe if a comet or asteroid the size of Tempel 1 ever came plummeting toward Earth. Scientists and world leaders alike got a wake-up call in that regard in 1994, when comet Shoemaker Levy-9 slammed into Jupiter and left scars visible from Earth for a year. Astronomers aren't sure whether it was an asteroid or a comet that collided with Earth sixty-five million years ago and wiped out the dinosaurs, so they study both as potential bombs that could cause humans' demise in a similar fashion.

One of the reasons for the uncertainty is that the difference between asteroids and comets is becoming less clear. It used to be pretty cut and dried. The story went that a giant cloud of gas and dust collapsed to create the solar system about 4.6 billion years ago. Asteroids formed from rock in the warmer, inner areas of the system, and comets from rock and ice in the colder, more distant regions. The first asteroid to be discovered was Ceres, in 1801, in the so-called asteroid belt between Mars and Jupiter. (As of 2006, Ceres has been reclassified as a dwarf planet by the International

Astronomical Union.) In the past two hundred years, a whole family of objects has been found in this region. Other than Ceres, they're all chunks of rock that are five hundred miles in diameter or less—too small to hold atmospheres but big enough to pack punches if they hit Earth. Even Ceres itself is barely wider than five hundred miles.

Astronomers think it was an asteroid that gouged a hole in the Earth just east of Flagstaff to form Meteor Crater. The term "meteor" throws a monkey wrench in any logical explanation of comets or asteroids, especially because meteors are tiny pieces of asteroids—but meteor showers are from the debris left behind by comets. The meteors that come from asteroids are typically tiny—fractions of inches—and they don't have to be much bigger than that to leave those fiery streaks in the sky. Most people attribute the phenomenon to Einstein's famous equation $E=mc^2$. They attribute a meteor's brightness to the idea that it takes just a tiny bit of mass to give off loads of energy as it converts to gas in the Earth's upper atmosphere. Schleicher did too, until he started doubting that reasoning and delved into the marvel a little more deeply. "What really makes these things bright given how small they are isn't actually the mass conversion but the friction of the particle with the atmosphere," he explains. "It ionizes the atmosphere and it's almost like a neon light. Most of it is really just the fact that the thing is going fast, lighting up the atmosphere. If it were really $E=mc^2$ it would be like having a small nuclear bomb, and that's not what we get."

Comets, historically, were defined by their tails. Because their wide orbits keep them for most of their lives in the frozen zone beyond Neptune, their surfaces are covered with dirty ice that vaporizes in fiery displays once they get closer to the Sun's heat.

The dazzling missiles were seen as ominous signs from God in ancient times, but these days they are welcomed as messengers from the Kuiper Belt or even the Oort Cloud, which is theorized to exist beyond it. Comets are typically much smaller than asteroids. The famed Hale-Bopp was thirty to fifty miles across. Halley boasted a ten-mile diameter. While many asteroids are big enough that gravity works to make them round, comets tend to have elongated shapes that get even more odd as interior patches of ice vaporize. Whereas asteroids' orbits keep them mostly within the inner reaches of the solar system, comets range far and wide—giving them lots of chances to meet with dramatic fates by going up against unyielding targets. Shoemaker-Levy 9, the comet that hurled itself into Jupiter, died in that violent crash. As Lowell comet expert David Schleicher points out, "Jupiter's doing fine, thank you." Some comets get flung into the Sun. Others come in contact not quite with a planet but with the formidable force of its gravity—and get tossed off their tracks altogether, moving onto haphazard trajectories through outer space.

Schleicher prefers it when comets orbit close enough to Earth to study—or at least visit every so often. Those are his bread and butter, he admits. It was Halley's anticipated arrival in 1986 that gave Schleicher a reason to come to Lowell in the first place. Hailing originally from Michigan, Schleicher had earned his doctoral degree in astronomy at the University of Maryland. By 1985, he had completed a post-doctoral program at Arizona State University in Phoenix, but his thesis advisor from Maryland, Mike A'Hearn, had already paved the way for Schleicher to take on his next project about one hundred forty miles north. A'Hearn—who would eventually lead the Deep Impact mission—collaborated for years with Bob Millis, whose career was dedicated to research until he became Lowell Observatory's director. The typical routine was for Millis to make observations from Flagstaff and to send the data to A'Hearn for processing. But Schleicher, having worked under A'Hearn, also knew how to process the data. So it was a natural fit for him to set up a post-doctoral fellowship at Lowell just as Halley was approaching.

When it buzzed Earth in 1910, Halley made a dazzling appearance that inspired all kinds of poems and songs. At age fifty-two, poet William Wilfred Campbell was near the end of his life when Halley visited. "Strange wanderer out of the deeps / Whence, journeying, come you?" he wrote in the opening stanza to his "Ode to Halley's Comet." "From what far, unsunned sleeps / Did fate foredoom you / Returning for ever again / Through the surgings of man / A flaming, awesome portent of dread / Down the centuries' span?" In 1990, country singer Mary Chapin Carpenter wrote a song called "When Halley Came to Jackson" about a baby whose father, holding her when Halley arrived in 1910, wished that his daughter would see it in 1986. And as recently as 2002, folk duo Michael Chevalier and Darwinfish, in their song "Halley," decried a friend whose "coming and going … singing the same old song" was reminiscent of that celestial visitor.

But Halley didn't sing its old song with quite as much gusto in 1986. And that was something of a disappointment for astronomers who were on the edges of their chairs for just as spectacular a display as in 1910. Any country with a spacecraft that year had it aimed at Halley. Russia and Japan each had two and Europe had one. The United States had planned one, but it got killed in budget wrangling. Instead, most of the country's ground-based telescopes, as well as most in the rest of the world, were trained on it. "Halley looked really wimpy, and it was not as spectacular a show purely because of where it was and where the Sun was," Schleicher said. "The vantage point was horrendous." He expects the view in 2061 will be pretty good, but it will still pale in comparison to 1910; and "the one after that is going to make 1910 look dull." Eventually, the Sun will burn off Halley's volatile components and it will cease to be a spectacular show not because of its position but because there's nothing left to vaporize. And therein lies one of the rubs in distinguishing asteroids and comets: "I expect Halley," Schliecher said, "to finally get to the point when it looks like an asteroid."

He says there might be plenty of objects that look like asteroids and orbit like asteroids, but really they are comets at the ends of their lives, when the volatiles have burned off or crusted over so they no longer yield the fiery tails. This blurring of the distinction is relatively new, based on a few phenomena in recent years that have revealed objects in the in-between stages. One of the most significant was the perplexing demise of the comet LINEAR, which disintegrated before the eyes of astronomers all over the world when it approached the Sun in the summer of 2000. Scientists think the solar heating triggered a massive disruption of the comet's fragile core—and Schleicher thinks LINEAR may have been primed to fall apart because it lacked much ice holding its core together. "We've now seen objects that sort of bridge the gap," he says, "not as much ice in comets, not as much rock in asteroids. Historically it was very clear-cut. Now we have to say, 'What's in a name?'" To finish out the name game, there's still the confusing fact that most meteors are tiny asteroids, but meteor showers come from comets. When material from the head of a comet burns off it first forms the fiery tail, and then it gets left behind in tiny fragments that are dust-size up to fist-size. The comet leaves enough of this castoff material that it gets distributed through its orbit. And if the

Earth's orbit crosses the comet's, it typically does so at the same time each year. This is why stargazers get to look for the Perseid meteor shower every summer and the Leonid meteor shower each fall. At the time of the meteor showers, the comets are elsewhere on their trip through the heavens, so there is no danger of collision— but Earthlings get treated to the displays they have left behind.

· · · · · · · · ○

For David Levy, a Tucson astronomer who started watching comets in 1965 and eventually helped discover twenty-one of them, Deep Impact was the reverse of a phenomenon he helped chronicle a little more than a decade before. Back then, Levy had joined forces to search for comets with Flagstaff astronomy couple Gene and Carolyn Shoemaker. Carolyn is still affiliated with Lowell Observatory and the U.S. Geological Survey in Flagstaff, and Gene's ashes went to the moon aboard NASA's Lunar Prospector in 1998 after he died in a 1997 car crash in Australia. A comet discovered by the trio, Shoemaker-Levy 9, shocked the world first because the comet got close enough in its orbit of Jupiter to be pulled into two dozen pieces—and then because the biggest of those pieces went hurling into the planet's surface, sending up plumes of debris and leaving extensive damage. The impact eventually helped fuel NASA's four million dollar annual project to find comets and asteroids that could collide with Earth. NASA's efforts, which got under way in earnest in 1998, built on a Spacewatch program that was already using two telescopes on Kitt Peak and coincided with the start of the Lowell Observatory Near-Earth Object Search (LONEOS). In 1999, congress gave NASA a mandate to find ninety percent of all the near-Earth asteroids greater than a kilometer across—or 0.62 of a mile—by the 2008. The mandate underscored the efforts already underway and infused them with a new sense of urgency. By that time the directors of each of those programs, Tom Gehrels at Spacewatch and Ted Bowell at LONEOS, were already a few steps ahead.

Gehrels started using a single telescope in the early 1980s atop Kitt Peak, fifty-five miles southwest of Tucson, to search for large asteroids that could hit us. He was more than a dozen years ahead of his time, so he mostly worked alone. "I needed nineteen million dollars, and almost nobody believed in asteroids, and almost nobody believed in the hazard," he says. But Gehrels' deep convictions drove him to seek corporate donors who would fund his vision—and Spacewatch was born. That first telescope was joined by a second one, four times as powerful, in 1997, just in time to serve NASA's mandate. Meanwhile, Bowell's career path was leading him to a similar place. Bowell earned a doctorate at the University of Paris and then arrived at Lowell. He spent his first few years studying the surfaces of planets using a technique called polarimetry, which measures the dispersion and other properties of light. One of the most interesting—and reassuring—things he discovered was that Apollo, when it landed on the moon, wouldn't sink and bury its astronauts in a fine-powder grave. But Bowell had become disillusioned with polarimetry, especially when other techniques came about that could analyze planetary surfaces more efficiently. Right when he was ready to change direction entirely, Bowell ran into Henry Giclas, who offered to take him out to Anderson Mesa and show him how to blink plates. Bowell loved it. He used those plates to discover myriad asteroids and other minor planets. At the time, there were only two thousand asteroids with known orbits—not even enough to figure out the structures of the asteroid belts. So Bowell devised a program to find new ones, as well as re-discover old asteroids that had been spotted throughout history and then lost. Soon, his rate of discovery "far exceeded what was going on anywhere else in the world, combined," he says. "At one point I became the leading discoverer of asteroids." Bowell kept it up through the late 1980s, by which point ten thousand asteroids had been documented with accurate orbits. Many of those were found at Lowell.

Bowell's program hit a rough patch then because Kodak stopped making the plates that were used to compare swaths of sky in search of his fleeting asteroids. "Who wanted to use eight-by-ten or eleven-by-seventeen plates anymore? No one," Bowell says. "The plates they did produce gradually got less sensitive." But Gene and Carolyn Shoemaker soon came to the rescue. By then their program to observe minor planets at Palomar Observatory was well underway, having gotten its start in 1983. And they were using film, not plates. The pair would cut out circles of film, bake them with gas to speed up the emulsion, load them into a holder, put it in the telescope, and instruct a person to open and close the shutter at an appointed time. One night of dedicated observing yielded about sixty films, which the observers would develop in a darkroom and scan under a microscope for moving objects.

At the time, Carolyn Shoemaker's background was in history and political science. She had spent one passionless year teaching those subjects. Gene Shoemaker was a geologist at the U.S. Geological Survey—an Earth geologist. David Levy, who became one of their most famous co-discoverers, was trained in English literature. "None of us were astronomers," Carolyn says. "None of us at the time had even taken an astronomy course." And yet, she'd found a thing she was excited about—and they were all good at it. Carolyn Shoemaker discovered thirty-two comets on the couple's films between 1983 and 1994. All of them bear the name Shoemaker, and many have hyphenated names because of the couple's colleagues. By the late 1980s, Bowell was in full collaboration with the Shoemakers. Their program infused his with fifty times the power it had when he was going it alone.

The Shoemakers' success was also their project's undoing. The Jupiter impact shocked NASA into wondering what would happen if a celestial body ever collided with Earth, and started the agency's push to find those natural missiles. Since 1998, NASA has spent thirty million dollars to fund the same five programs for that work: Spacewatch in Tucson; the Lincoln Near Earth Asteroid Research (LINEAR) project, operated by MIT Lincoln Labs near Socorro; Tucson's Catalina Sky Survey; LONEOS; and the Jet Propulsion Laboratory's Near-Earth Asteroid Tracking program.

All the programs are using technologies that to Shoemaker seem "magical": fully-automated scopes that can be controlled from distant computers and CCD cameras that can capture anything in range that appears to move. Even amateur astronomers, unless they are into astrophotography, are using scopes equipped with CCD cameras. All by itself, New Mexico's program, LINEAR, can get twenty thousand asteroid positions plus some comets in a single observation period, which usually lasts about ten days. The Catalina Sky Survey is even more powerful, following upgrades to its telescope last year. So far, the global search has documented eleven hundred large asteroids near Earth. The pace of discovery hit a peak in 2000 and 2001 but has been slowing because there are fewer asteroids to find now. Astronomers have also discovered more than three thousand smaller asteroids that may not cause environmental damage akin to a nuclear winter, the way the larger ones might, but that would be risky just the same. Even asteroids six hundred feet across, "if they were to hit a city, would be devastating," said Stephen Larsen, director of the Catalina Sky Survey.

The community of professional astronomers involved in such searches is actually rather small. At last count, NASA reported that there are about one hundred. But Larson said they get help from a broad network of amateur astronomers more than happy to track sightings by the pros. Astronomers of both types use a Web site maintained by the Minor Planet Center in Cambridge, Massachusetts. The professionals document the asteroids they find, and then the amateurs can follow those and add their own sightings to the database. No deadly asteroids have been confirmed to date—that is, none is headed this way in the foreseeable future. But Larson and others believe it is not too soon to figure out what we would do if astronomers did unveil one. "So far, there's been no funding to decide how you'd deflect it," he said, adding the searches are a vital first step to any such efforts. "If you can discover these things fifty years ahead, you have a chance." Meanwhile, astronomers have made some intriguing discoveries, including a couple of objects to watch with one eyebrow raised.

In July of 2001, a Lowell observer and Northern Arizona University graduate student named Michael Van Ness discovered 2001 OG108, one of only fourteen known asteroids belonging to a rare class called Damocloids. Named after (5335) Damocles, the first asteroid found of its type, Damocloids have orbits something like comet Halley's—highly elongated tracks that reach as far as the Oort Cloud up to one hundred times farther from the Sun than Earth. As such, astronomers think Damocloids could be the remnants of comets. Even among that class, 2001 OG108 was an unusual find because it was the second of only two such objects with orbits that, on the close end, bring them closer to the

Sun than Earth. Both such objects were discovered by LONEOS. Fortunately, 2001 OG108 poses no near-term threat of collision with Earth. The closest it can come to our planet is about thirty million miles. Perhaps of significance for the long-term evolution of its orbit is that the asteroid can approach Jupiter within one hundred million miles. And if Jupiter has its way with OG108, there is no telling what might happen.

In September 2003, LONEOS documented a house-size asteroid that came within eighty-eight thousand kilometers of Earth. That is less than a quarter of the distance from the Earth to the moon, making it the closest well-documented flyby of an asteroid that did not breach the atmosphere. It was an unusual find for LONEOS because it was so small. "In a good month, we find five to ten near-Earth asteroids, but usually, the ones we discover are as big as mountains, or at least football stadiums, so this one was unique for us," Ted Bowell said at the time. Usually, even today's CCD cameras are not quick enough to snap an image of such a small body. Known as 2003 SQ_{222}, the asteroid was imaged a few hours after close approach by Van Ness who was still an observer, and still an anthropology graduate student at Northern Arizona University. SQ_{222}'s known brightness and distance allow calculation of its size—three to six meters—making it one of the smallest asteroids for which we have a reliable orbit. If SQ_{222} had approached Earth, it would have exploded harmlessly in the upper atmosphere with an energy comparable to that of a small atomic bomb. "Objects the size of 2003 SQ_{222} actually do burn up in Earth's atmosphere every year or so, producing a spectacular light show," Bowell said. In what is most likely a coincidence, an intense shower of meteorites was reported in India about ten hours before 2003 SQ_{222}'s closest approach to Earth; astronomers did some research to test the unlikely possibility that the asteroid and the meteorites could have been fragments of a larger asteroid broken apart by a celestial collision, or disruption during a previous Earth approach.

The very next year, long-time Lowell observer Brian Skiff re-discovered Hermes, a near-Earth asteroid that had been seen and then lost sixty-six years earlier. That would have been cause for excitement all by itself, but Skiff also noticed the asteroid was unusually bright. Sure enough, astronomers following his lead saw that it was not one but two objects, a binary pair circling around each other while they approached Earth. By November it came within four million miles of Earth—close enough for amateurs to see with backyard telescopes. At the time, Bowell called it "the Holy Grail of near-Earth asteroid discovery ... an asteroid's return just does not become more profound than this." It just so happened that on the same day Skiff captured those images, Discovery Communications, Inc. and Lowell Observatory announced a partnership to build the new Discovery Channel Telescope at Happy Jack, a site southwest of Flagstaff and not too far from the LONEOS hub at Anderson Mesa. One research goal for the forty million dollar telescope is to expand the search for near-Earth objects, including those smaller than Hermes.

· · · · · · · ·

If it is going to keep up with the search for near-Earth asteroids, Lowell's program will need a boost. So will every other program currently being funded by NASA for such work because there is a new crop of telescopes in the planning or early building stages that will push the hunt to a drastically higher level. The race is on partly as a result of a new, revised mandate by NASA in 2003. Following the agency's "Study to Determine the Feasibility of Extending the

Search for Near-Earth Asteroids to Smaller Limiting Diameters," the new threshold for asteroids and comets that NASA worries about has been lowered to just four hundred twenty feet across. Congress agreed in 2005, ordering NASA to find all such asteroids within fifteen years. But objects that small are hundreds of times harder to find than those above the one-kilometer threshold— and nearly impossible to find with a telescope the size of the one LONEOS uses: "LONEOS wouldn't make a dent in that," Bowell says. Instead, four new, multi-million dollar telescopes are lining up to try.

The first two, the University of Hawaii's Panoramic Survey Telescope & Rapid Response System (Pan-STARRS) and the Discovery Channel Telescocope at Lowell, would together accomplish NASA's new goal within ten years. Both are partly funded and construction is under way, although as of summer 2006, the University of Hawaii was still choosing a site. There are slight differences between the two telescopes. Pan-STARRS will comprise four 1.8-meter (six foot) mirrors with a combined field of view of three degrees. The DCT will have a single, 4.3-meter (fourteen foot) mirror with a field of view of two degrees. So it would take longer to image the entire sky, but the optics would let in more light, for an ability to see fainter objects in space. But the DCT's ability to be that powerful, and to be a player in the next generation NEO search, will rely on Lowell Observatory's ability to collect funds for the wide-field camera.

"If DCT were the first telescope to come online with a wide-field camera, it would rule the sky," Bowell says. That is, at least until the Large Synoptic Survey Telescope, or LSST, comes aboard. That telescope is still being conceptualized at the University of Arizona and got a huge boost in early 2005, when a private donor and astronomy buff gave the university $2.3 million toward the project. The 27-foot-mirror telescope is slated to go to Chile, where it will spy on near-Earth asteroids of all sizes and the rapid changes that occur with exploding stars. By mapping the positions of stars, planets, and other celestial bodies, the scope is also expected to reveal more about the action of so-called dark energy—the repulsive force between space objects first proposed by Albert Einstein—that keeps the universe from collapsing in on itself from its own gravity. And LSST would not need the help of any other telescopes; it could accomplish NASA's NEO goal alone within ten years.

The fourth and final contestant, for now, is the Near Earth Object Camera, or NEOcam, a space-based infrared telescope system proposed by NASA's Jet Propulsion Laboratory in California. Like the LSST, fans of the NEOcam say it could do the whole job by itself in ten years' time.

Astronomers are starting to point out, however, that finding Earth's celestial enemies is just the first step toward averting harm by them. So on another front, they are working to plot more ways to attack such objects before they reach us. Deep Impact was just the beginning. At an unpublicized 2006 meeting of the minds in Vail, Colorado, select astronomers gathered alongside mysterious men in black—probably weapons types, according to some who were in attendance—to start brainstorming ways to make even deeper impacts. The Tempel 1 experiment, after all, only left a football field-size pockmark on a four-mile-wide potato.

Part of the equation is nailing down the composition of comets and asteroids. The world has seen plenty of images of comets' tails, and previous missions have photographed the outer layers of the comets Halley, Borelly, and Wild 2. But astronomers think the real clues about the structure and makeup of comets and asteroids lie a few feet beneath that outer crust. There, they think, comets resemble the planets during their infancies and are therefore most interesting to study. And therein also lies the Achilles heels that will enable us to destroy them.

Chapter 8

TELESCOPE WARS

The U.S. Naval Observatory's Navy Prototype Optical Interferometer, on Lowell Observatory land at Anderson Mesa, would have been an ambitious undertaking no matter who in the world had built it. Sprawled over eight acres, the interferometer is designed to combine beams of incoming starlight from six mirrors with exquisite synchronicity, simulating the viewing power of a single telescope with a mirror a quarter-mile in diameter. At most, engineers had tried to coordinate up to three mirrors in the past, with tenuous success. Lowell Astronomer Nat White works on the NPOI with two other project managers, one each from the Naval Research Lab and the U. S. Naval Observatory. White coordinates day-to-day operations of the telescope and directs a staff of nine people who maintain it, observe the computer images it generates, and keep the mirrors clean. Don Hutter, the USNO's project manager, is responsible to communicate with a board of directors about progress on the behemoth telescope's ongoing construction and operation.

White was a part of a team that oversaw the coordination of three mirrors in the summer of 1994 and pulled all six together early in 2000. White was elated. He called the local news media and touted the telescope as one of the biggest and best in the world. He said pulling all six telescope mirrors together would allow for "imaging that no other telescope can manage at this time." That's because the six scopes, working in combination, have the power to provide fifteen telescope-to-telescope baselines, or one-dimensional views, of whatever object they are pointed at. When combined, they can reveal the most detailed views astronomers have ever seen. Furthermore, it boasts an experimental network of pipes that transport light from each mirror in a vacuum. That means the NPOI can utilize the optical wavelengths—that is, visible light. All other existing interferometers work in the infrared to radio wavelengths, which can hold up under less pristine conditions. The field of view for an interferometer is very small—in most cases, the width of a single star or several distant stars that are very close together. But it sees that star with phenomenal resolution. The NPOI showed early success, and scientists celebrated the first fuzzy images when all six telescopes were brought online. They cheered even louder when it observed the triple star system Eta Virginis, located about one hundred thirty light-years away from Earth, with unprecedented resolution. At the time, that success was enough for White.

"We're not looking at stars outside of our galaxy; we're not even looking at stars at the other side of our galaxy," he said. "We're just looking at stars close to us—but there are thousands of stars that we can look at with this high resolution. That's where we will be able to shine ... pardon the pun." But the engineers were not done yet. The interferometer would be complete, White thought, when each of the six telescopes had been relocated to eight hundred thirty-three feet from the center of the complex, at a distance of fourteen hundred feet from each other. With respect to resolving star details, he predicted that astronomers would be able to do the equivalent of standing at Upper Lake Mary, southeast of Flagstaff, and inspecting the eyelash of a mouse twenty miles away on Humphrey's Peak. It will be like seeing the dimples on a golf ball one thousand miles away, he said at the unveiling, or like inspecting a dime in Washington, D.C. from Flagstaff.

"It will lead to the direct imaging of the surfaces of stars and of star spots, analogous to the sunspots on the sun," added Kenneth Johnston, scientific director of the Naval Observatory. "This technology can also be applied to space systems for remote sensing of the Earth and other objects in the solar system, as well as stars and galaxies." So the interferometer would complement the work of survey telescopes like the Hubble Space Telescope and the Discovery Channel Telescope by honing in on objects of interest that they find. The master plan was to use it for astrometry—the precise measurement of star positions that takes into account their motion relative to Earth—which has become solely the purview of the USNO in the years since Giclas compiled his proper motion catalog. Refining and building on star maps could enhance Global Positioning Systems and allow for better-directed space flight. Overall, the interferometer was designed to help scientists complete detailed maps of the heavens, and it still could. But five years after that celebrated unveiling, it's been mired in so many technical challenges that it hasn't done so yet.

Any interferometer works on the principle that two coinciding waves of any type—whether visible light, infrared radiation, or even particles such as atoms that can act like waves—will amplify each other, while two waves in opposite phases will cancel each other out. In astronomy, an interferometer is good for cancelling out interference from a point source of light, like a star, and giving astronomers a razor-sharp image of the target object. The separated mirrors combine to achieve the effect of a single telescope mirror the size of the distance between them. The NPOI, when developed to full capability, will act like a single telescope mirror a quarter-mile across. If you added a lot more telescopes to the design—indeed, if you filled the entire, eight-acre site with telescope mirrors—you would get a complete image. "We can't do that. That would be impractical," White says. Instead, the six mirrors start with skeleton views of the objects they are programmed to study. The Earth rotates a little bit, and they capture a different view. That is like adding six more mirrors. And when they do that over and over, it mimics the effect of filling the whole site with mirrors. "Eventually," White says, "we build up an image." So the span of the whole apparatus determines the field and depth of view. The idea is that the NPOI's long baselines can help it see wide swaths of the sky in extraordinary detail, down to very faint and distant objects.

However, interferometry is hard. It is technically challenging to recombine wavelengths that have been split apart and line them up precisely enough to find the places where they agree. Lu Rarogiewicz once operated the Navy's Mark III interferometer —the house-size prototype—that was dismantled on Mount Wilson in the late 1990s, that the NPOI replaced. He brought this difficulty home in a 2001 description on Space.com of the conceptual Michelson interferometer, invented by American physicist Albert Michelson, who lived from 1853 to 1931. That interferometer design combines light beams from two flat mirrors. "Michelson's early experiments were affected by street traffic vibrations up to 1,000 feet away!" Rarogiewicz wrote.

White marvels at how it has taken one hundred years for the technology to catch up to Michelson's idea. Radio interferometry was developed into a practical tool during the 1950s and 1960s. The method was extended to measurements using separated telescopes by 1974 in the infrared and a year later in the visible wavelengths. The red giant star Betelgeuse was one of the first targets for diameter measurements using optical interferometry. Finally, computer processing advances in the late part of that decade allowed for interferometers fast enough to overcome the blurring effects of the atmosphere.

In the 1980s, interferometry evolved enough to start imaging the surfaces of stars, and that capability has been put in use at the Infrared Optical Telescope Array at the Fred Lawrence Whipple Observatory, operated by the Smithsonian Astronomical Observatory in Amado, Arizona. Interferometry is also being incorporated at the European Space Agency's four-telescope Very Large Telescope Interferometer, the six-telescope Center for High Angular Resolution Astronomy (CHARA) of Georgia State University, and the ten-telescope Magdalena Ridge Observatory Interferometer (MROI), slated for the Magdalena Ridge Observatory near Socorro, New Mexico. The first two are at least partially operational, but the latter is much earlier in the construction process.

The Navy started drafting plans for such an instrument in the 1970s, and it came up with a model about the size of a table. That was called Mark I. The Mark II, in the 1980s, was built to fill a room.

The 1990s model was about three hundred feet across, called the Mark III, and it was used to get the diameter of about one hundred stars from its perch at Mount Wilson. "When the Navy decided to build this one, it went for the largest interferometer that could exist under Earth's atmosphere," White says, referring to the NPOI. But they needed a place to put it. "For a number of years they looked for a site to build a very large interferometer in California, and they couldn't find one. A friend and I had written a paper in 1989 about how Anderson Mesa would be a good site … and the Navy got wind of it."

A Memorandum of Understanding was signed in 1991. Lowell would be a partner mostly in the sense that it would lend its space on Anderson Mesa for the new, cutting-edge instrument. But then, the Navy started looking at the costs of building such an instrument, and they were "astronomical," White says, "so Lowell became the contractor to build the infrastructure; that's when I joined the project." The interferometer is very much a working, fully developed instrument. But it is still fair to dub the NPOI a prototype because it is pushing the technology further than anyone thought it could go. "When we started this telescope, even when it saw first light four or five years later, it was way ahead of other interferometers in terms of how many telescopes we could combine," White said. "Nobody still has duplicated six mirrors." And the NPOI is one of only two interferometers in the world that works in the visible light. Most others work in infrared wavelengths, which are more forgiving of incongruities when they reach the central hub of a multiple-mirror telescope. Only one other interferometer, the Sydney University Stellar Interferometer in Australia, uses visible light. But SUSI uses only a single baseline—two telescopes—compared to the NPOI's six.

The builders of the NPOI knew they were up for a challenge when they tried to combine visible light beams from six distant mirrors. They started with an infrastructure of pipes linking all six mirrors across the eight-acre complex to a central hub where the beams must be perfectly combined. Just going through the air would disturb the beams, so the pipes are impeccably clean and a vacuum system draws all the air out that could interfere with the light's motion. It works because each of the three, eight hundred thirty-three-foot arms—all the arms can support up to three of the telescopes, depending on the configurations that are being used—includes three sets of eight-inch aluminum pipes. Each one leads from a telescope position to the instrument's central hub. And each one comprises twenty-foot sections that are adhered to each other with Neoprene bands and hose clamps. Inside the optical building and on each arm are sets of vacuum pumps that suck air out of the pipes. Then the pipes are sealed and the vacuum pumps turned off; otherwise, their vibrations would disrupt the images. The pipes are kept at one hundred thousand times less than atmospheric pressure. When starlight does hit a telescope mirror, it is directed by more mirrors through the pipes and into the optical building. Those mirrors move back and forth in the vacuum to compensate for the Earth's motion and the shimmering of the atmosphere. About one hundred fifty computer micro-processors respond to the tiniest vibration or temperature changes —caused by so much as a shadow passing over the pipes. To form an accurate image, all the incoming light beams must reach the optical building within a few millionths of an inch of each other. Heat would destroy the synchronicity, so cooling mechanisms control the temperature to within one-one hundredth of a degree. Don Hutter says no one has crunched the numbers to figure out how much the Navy has actually paid for the NPOI. "It's probably in the range of forty million dollars to fifty million, but no one knows," he said, adding that the labor is the factor that's been hardest to quantify. And now, fifteen years after ground was broken for the instrument, White admits the project founders may have bitten off more than they could chew. "The hope in the very beginning was that the instrument would reach its potential very quickly," White says now. "Maybe the Navy was overly optimistic ... and maybe Lowell was naïve."

First, the interferometer ran into unpredicted motions—some of them from ambiguous sources—that interfered with the ability to match up the light beams. So in the early 1990s, project managers found themselves hunting for lasers they could send through the pipes to measure the disturbances. But in the 1990s, such lasers were not commercially available. The only ones suitable for the job were being manufactured by private individuals, and they were pricey—up to six thousand dollars apiece. White says the difficulty of getting stable lasers illustrates that the technique in the visible region was pushing available technology, including stable lasers and fast, inexpensive computers. Second, getting the NPOI to achieve its maximun potential has been like peeling the layers of an onion: so far, solving each problem has inveiled another, more subtle problem. The new lasers were supposed to measure the minute irregularities caused by everything from miniscule temperature differences to the vibrations from mystery sources—presumably faint breezes or the rumbling of trains passing through the heart of Flagstaff fifteen miles away. Once the lasers nailed down that information, White and his colleagues thought, it would give astronomers a corrective factor they could use to balance out the irregularities and salvage their images. It didn't happen that fast.

"We took off that layer and below that there was another variation," White said. "We're tracking that down now. Now, we're approaching advertised precision." That's especially true for the part of the interferometer's job that has to do with astrometry, or measuring star positions.

As it stands now, the interferometer has been booked for research every night since the first couple of baselines started operating in the late 1990s. It has yielded numerous publications in prestigious journals such as *Nature* and the *Astrophysical Journal*. Using it, researchers have discovered that apparently single stars were actually closely linked binaries, like Zeta Orionus, the bright star marking the eastern end of the "belt" on the constellation Orion. They have measured their short orbital periods, sometimes only months long. They have watched the light change as stars rotate, and they have constructed theories about what that means for the physics of those stars.

But there was supposed to be much more. The NPOI is the only interferometer in the world trying to do astrometry, or make star catalogs, over the whole sky. At peak performance, the instrument is designed to map a few hundred stars in a few years. Half the telescope's time was supposed to go toward astrometric work of measuring the position of stars in the sky like cities on a map. The positions would have been one hundred times more accurate than any previous measures using Earth-based telescopes. And it would have followed up on star maps constructed by Hipparcos in the 1980s. Those maps were cutting edge at the time but have fallen into obsolescence in some cases because of the proper motions of some of the stars. A telescope such as the interferometer could correct those discrepancies. And it could track the same stars over and over, keeping the maps current on a near-constant basis.

It would also be helpful for work on more distant binary star pairs. Otto Franz and a colleague, Lisa Prato, tried to use the NPOI for a project on star pairs but had to quit, because the telescope just could not see faint enough.

That type of work awaits the next fix for the interferometer, at a projected cost of about five hundred thousand dollars: high-tech optics called beam compressors that will correct a limitation caused by the pipes. Each of the telescope mirrors is twenty inches across, but the pipes allow for six-inch-wide beams. The additional optics will take the light and compress it, letting in four times as much light. There is a chance that the compression will cause another layer of trouble with the images because compression tends to magnify any irregularities in the light hitting the telescopes. But if it works—if the optical waves hitting the mirrors are fairly uniform—the difference will be profound, allowing astronomers to see images fifty to sixty times fainter than they can see now and lending greater contrast to all the NPOI's images. "I hate to guess the future with the interferometer," White says, "but I expect within eighteen months or so we should know whether that instrument is going to be useful or not to the general astronomical community."

Meanwhile, other telescopes are beginning to catch up to the NPOI. Europe's Very Large Telescope Interferometer is one of an emerging class of "outrigger telescopes," which combine main mirrors with moveable auxiliary telescopes. The VLTI will use four main scopes and four six-foot auxiliary mirrors. "Where they have us beat is faintness," White says. "But they're not working in a vacuum system, so they're limited to the infrared." The NPOI's optical niche allows it to contribute the type of seeing to space science that adds depth to our understanding of the atmospheres of stars, penetrating more deeply than observations in the infrared or any other region of the spectrum. So, "our unique niche is still there, but bits and pieces of it are being chipped away," White says.

Overall, he doesn't know the prognosis for the future of interferometry. There is still an interest, which is evident in the number of such instruments currently in the works. But so far, he says, they have all taken longer to build than expected—and have produced poorer results. "From my point of view, the jury is still out on them," he said.

For sure, an instrument like the NPOI will be exceedingly advantageous if it can be used in concert with other technologies. Radio astronomy is a common companion field for interferometry partly because radio waves are so long that they can be recorded separately, timed, and combined later. And its detail-oriented approach would be the perfect complement to the swallow-the-sky-whole capability of a survey instrument like the Discovery Channel Telescope, expected to see first light at Lowell Observatory as early as 2009. That is because the interferometer allows a detailed look at the map of a star's surface. It may show spots or blotches and the shape of the star related to its rotation around its axis. If at the same time the DCT is aimed at that star with a high-resolution spectrograph, astronomers will also know the state the various molecules and atoms are in, its temperatures, pressures, turbulence, polarization, and abundances of its various elements. "A rough analogy," White says, "is a map of the Earth by the interferometer and then add the geology with the DCT."

Chapter 9

FOR ALL TO SEE

Even as Nat White struggles to get the interferometer working at its full potential, Lowell has embarked on another project it hopes will secure its place in the world of competitive observatories.

The Discovery Channel Telescope (DCT) is expected to cost forty million dollars to build, with a target completion date of 2010—a date that has been inching into the future since the project began. Still, if it is finished by then, the DCT will become the fifth largest telescope in the continental United States. It is a huge undertaking, with a fourteen-foot, sixty-seven hundred-pound mirror inside a massive, three million-dollar dome. The DCT is conceived as a survey telescope, a "general workhorse," says Project Manager Byron Smith. Astronomers want to command it to do everything from spotting asteroids on trajectories toward Earth to chronicling the birth of galaxies in faraway parts of the universe. If the funds come in and building stays on schedule, the Discovery Channel Telescope will, for several years, lead the emerging pack of telescopes useful for such efforts. Lowell astronomers say the telescope will assure the observatory's future. And the whole project, from the summer 2005 groundbreaking all the way up to its research feats, is eligible for broadcast to the Discovery Channel's worldwide audience of four billion people. The station is also planning a two-hour movie called "The Making of the Discovery Telescope." This marks the first time a television company has funded so much of a telescope—a full third of the money. The innovative partnership got its start in 2003, when founder, president, and then-CEO of Discovery Communications, John Hendricks, was serving on the Lowell Observatory Advisory Board. Hendricks touts the unprecedented partnership as an innovative combination of research and education. "The Discovery Channel Telescope will make one hundred or one hundred fifty discoveries each viewing evening, if you can imagine that," he said. "Some of those will be so much fun to be able to transmit into the nation's classrooms, to build shows that we can air worldwide."

· · · · · · · · ○

So the partnership is a feather in the cap of The Discovery Channel, Flagstaff, and Lowell Observatory—in more ways than one. In large part, it will greatly enhance Lowell's seeing power, and keep it in the big leagues of astronomical observatories. The telescope mirror alone, at fourteen feet, more than doubles the size of the mirror of the next biggest telescope: the 1960s, 72-inch Perkins telescope on Anderson Mesa. And it's about time, Millis says. "There has been a real burst in the construction of larger and technologically more capable telescopes," he said. "And so acquiring such a telescope is important purely from the standpoint of competitiveness."

Acquiring such a telescope also meant months of sleepless nights when Byron Smith first took the helm. Smith earned his master's degree in mechanical engineering from the University of Texas in 1996 and went to work for VertexRSI, a company that at that time made "big things that point precisely at the sky": radio and optical telescopes and radar devices. "Pretty much everything they did was the size of a house or bigger." The company has since expanded into the small portable antenna market. While he was there, Smith cut his teeth on a radar ballistics missile defense program and several satellite communication antennas. He had lesser involvement with the Southern Observatory for Astrophysical Research (SOAR) and the Visible and Infrared Survey Telescope for Astronomy (VISTA) telescopes, both in Chile.

Meanwhile, in Flagstaff, a man named Thomas Sebring was assembling a project team to manage construction of the Discovery Channel Telescope, "most of whom have now, unfortunately, departed with the exception of Byron Smith who I hired away from VertexRSI," Sebring said. Sebring was hired based on a track record of managing the design and construction of projects including the Hobby-Eberly Telescope in west Texas and the SOAR telescope. He managed the design and construction of the DCT headquarters building on Mars Hill, as well as the development of a complete concept design for the telescope. He orchestrated a peer review with a panel including astronomers and telescope project managers all over the country. The designs passed with flying colors, but the funding wasn't going so well. "When I arrived at Lowell they had succeeded in raising twenty million dollars toward an estimated cost of fifty million dollars for the telescope including the wide field of view camera. In spite of concerted efforts by Bob Millis and myself, we did not succeed in raising any additional funds." The men visited universities, made presentations to the National Science Foundation, NASA, the U.S. National Optical Astronomical Observatories, and the U.S. Naval Observatory. They worked to identify individuals who might make charitable contributions. They met with Arizona congressional representatives and made plans to lobby the federal government with the help of major potential contractors and the Discovery Channel. "Had we been successful," Sebring says, "I have every reason to expect ... that we would have reached first light in 2007 and handed the telescope over to full science operations in 2008 as planned."

Sebring did not stick around that long. In the spring of 2004, just a year and a half after agreeing to oversee construction of the DCT, Sebring was offered a job as project manager for the Cornell Caltech Atacama Telescope in Ithaca, New York. He gave notice, recommended Smith to take over, and was gone. So Smith stepped in as the person charged with overseeing the telescope's design and construction, as well as coordinating its capabilities with the goals of the Lowell astronomers, which are always evolving. It's a big job, the stakes are high, and there are a lot of moving parts.

For starters, the telescope has to be diverse enough to accommodate twenty doctoral degree-holding astronomers on Mars Hill, as well as visiting researchers. It has to be strong enough to accommodate a two-ton camera hanging off the telescope tube. And it has to be modern enough to incorporate loads of cutting-edge technology, some of it being built for the first time, without a hitch. Telescope mounts, for example, have progressed beyond the two-axis equatorial mounts that were once standard. Instead, engineers are opting for a design that allows the base to rotate as well. The new mounts are more compact and cheaper, but it's an altogether more complicated task to coordinate motion in three directions. The field of optics is also evolving. When the DCT mirror was cast and fused by Corning, Inc. in Canton, New York, it "was proclaimed the most perfect one ever made at the Canton plant—much to the relief and excitement of the highly specialized crew that created it," wrote Molly Hermann, a producer for Discovery Channel, in a project update. Made of Corning's ultra-low-expansion glass, the mirror is only four inches thick. The materials let heat dissipate faster so it does not distort the telescope's seeing—but they are flimsier. In the fall of 2006, the University of Arizona's (UA) optical sciences engineering team started bonding a minimum of one hundred twenty pucks to the mirror's convex backside to make a support structure for holding the mirror just as it will be held in the telescope. The system will ensure the mirror does not flex under the force of grinding and polishing. Grinding the mirror to get it closer to the ideal shape will take about five months, according to Martin J. Valente, director of the university's optical fabrication and engineering facility and UA's principal investigator on the project. Polishing the mirror and figuring it—the final stage that will make the mirror accurate to within a fraction of a wavelength of light, or a few millionths of an inch—will probably take another fifteen to eighteen months.

The part of the design that has Lowell astronomers on the edges of their seats is the wide-field camera. Goodrich Optical and Space Systems in Danbury, Connecticut drafted conceptual design studies for what Smith believes is the biggest telescope camera that has been built to date—and they are likely to bid on its construction if Lowell secures the funds. The entrance lens that would be visible from the outside would be nearly four feet across, with six lenses behind it half that wide. The idea is to produce an amazing image across a wide field of view. "In a single exposure it can image a circular region of the sky whose diameter is equivalent to four full moons lined up edge to edge," Millis boasted at the 2005 dedication. The design includes a three hundred-megapixel camera. (Compare that to the six-megapixel point-and-shoot models popular with amateur photographers.) The camera requires its own cooling system so heat does not interfere with its images. It will contain a filter changer, so astronomers can swap out colored pieces of glass to determine the colors of objects in space—a clue to characteristics including temperature, size, and age. And it will incorporate a state-of-the-art atmospheric dispersion corrector to counteract the weak prismatic effects of the Earth's atmosphere.

Alongside the engineering feats he has to oversee, Smith must also coordinate construction elements with widely varying timelines. The DCT facility—including an auxiliary building where the mirror will get its annual refinishing—was well under way by the summer of 2006 and expected to be done the next year. The next step is for someone to come on board to start developing the computer software that will run the behemoth instrument. Just writing that could take years. "Historically it's taken ten years of effort to develop the software necessary to develop a telescope like this. But a lot of the components people once had to write by hand, you can now go buy a commercially supported package," Smith said. Still, that part of the project will take several years at least. Modern telescopes rely heavily on computers, and that reliance means the computers and the instrument have to be in synch at all times. "There's hardly any part of the telescope that doesn't have a computer associated with it," Smith says. "Ongoing maintenance will be a challenge." So Smith has spent a lot of time visiting other telescope sites to see what works. And he has been watching the places where other designers fail so he can avoid repeating their mistakes. A prime example is the SOAR telescope in Chile. The mirror is the same size as what is slated to go into the DCT, so Smith used that design as a starting point. But that telescope suffered a huge setback when the mirror's lateral supports—the devices meant to keep the mirror from falling sideways as the telescope aims for the horizon—caused the mirror to bend and blur the images it produced. The problem was a result of a mistake in the design phase, Smith said—part of his job with DCT. At SOAR the mistake caused a two-year delay in the science and cost several million dollars. "It's humbling," he said. "I've met the people involved. There are a lot of smart people and you see how they got into it." Smith said for the first few months of his job, he would frequently wake up in the middle of the night "in a startle." He says he has gradually gotten more relaxed: "We're probably screwing something up. But that's part of building a telescope."

Besides, Smith and his team at Lowell are going to great lengths to make sure they do not screw anything up. The result of the 2004 design review was reassuring—the reviewers found no obvious errors. The project will undergo preliminary design review in mid-2007, which will be another welcome chance for oversight. So far, Smith has found the feedback a refreshing departure from the commercial world and its trade secrets that he left behind. "The astronomical world is more collegial," he said.

That's not to say it isn't competitive.

The funding uncertainties plaguing the DCT's early years are putting it in a head-to-head race with two other major telescopes that are already under construction, and one more that is still being conceptualized. The likely competitors are the Large Synoptic Segmented Telescope (LSST) out of the University of Arizona and the Pan-STARRS, a project of the University of Hawaii. Both are expected to be more costly than the DCT. The LSST will combine three mirrors to add up to the equivalent of an 8.4-meter (28-foot) mirror and is slated to go to a site in Chile. That scope would be dedicated to peering at Near-Earth Objects, but as a side project it could image stars. Pan-STARRS would comprise four 6-foot telescopes to simulate the power of a 15-foot telescope on the Hawaii mountain site of Mauna Kea. It is being designed with a field of view of three degrees, compared to the DCT's two degrees. That means Pan-STARRS could observe a wider swath of the sky in a single sweep than the DCT—but the DCT will see fainter. So the DCT would capture dimmer and more distant stars, light from younger galaxies, and smaller incoming asteroids. That is, if Lowell finds a partner to fund the ten million-dollar wide-field camera. Without it, the telescope will still be able to do many of the feats anticipated by Lowell astronomers. It will still be useful for infrared observations and spectroscopy of faint objects. It will still be able to show astronomers the distances to faraway stars, their temperatures, compositions, and changes in brightness—a useful parameter for work Lowell's Ted Dunham does, for example, in searching for extrasolar planets by the transit method. It will do all of those things and more, but not nearly as impressively as it would with its ten million-dollar companion camera. "If you have a big field of view you can do those things ten times more quickly," Smith said.

The beauty of the Discovery Channel Scope will be in its adaptability, Millis says. Not all the astronomers want to use the wide-field camera for their work. And if that was its only mode, the DCT would be crippled by the light of the moon—like LSST and Pan-STARRS will be. With its infrared and spectroscopy capabilities, the DCT can work even on moonlit nights. Overall, the DCT is slated to work in three main arenas: the history and future of the solar system, including an expanded Deep Elliptical Survey and NEO search and infrared investigations of trans-Neptunian objects and near-Earth asteroids. It's taken astronomers fifteen years to discover about one thousand trans-Neptunian objects, Millis points out. "With the DCT, we could do that in two nights of observing." The other two categories of work include the form, function, and evolution of stellar and planetary systems and the formation and evolution of galaxies. As of late 2006, the specific areas of study within those broad topics are still being determined.

· · · · · · · · ·

The Discovery Channel Telescope clearly comes at a time when the technology at Lowell needs a shot in the arm. Millis notes that the current research staff is good at getting grants to work on other telescopes, including the Spitzer and the Hubble space telescopes and leagues of them on the ground. But that alone qualifies Lowell as a research institution, not an observatory. He said Lowell's future came up during talks surrounding the institution's centennial in 1999. "Naturally, you start asking about the next one hundred years," he said. "Do we want to continue to be an observatory? We could be a research institution … or a science center, and augment the public programs." The consensus was that Lowell staff wanted to continue to work for an observatory—a renowned one. And that left them wanting more than what they had. Mars Hill still houses the original, 1896, 24-inch Clark Refracting Telescope used by V.M. Slipher, but it's being used as a public education tool, not a research instrument. Same story with the 13-inch Pluto Discovery Telescope used by Clyde Tombaugh. Lowell operates four research telescopes on Anderson Mesa, including the 72-inch Perkins, the 42-inch John S. Hall Telescope, the 31-inch National Undergraduate Research Observatory telescope, often used by Northern Arizona University observers, and the NPOI. Millis said all of them except for the NPOI have been "gussied up" to their physical limits. "If all we had downstream were the existing telescopes on Anderson Mesa, it would be hard to claim we were a significant observatory," Millis said. Lowell's first telescopes were powerful enough for the solar system work—including the discovery of Pluto—that earned Lowell professional respect for the first one hundred years. But they are not powerful enough for the next step: the detailed study of fainter and farther objects in space. So the DCT isn't going to make or break Lowell the institution, Millis says—but it is expected to guarantee the future of Lowell Observatory.

Epilogue

DARK OF NIGHT

The lights of cars creeping along the two-lane reservation road far below Kitt Peak can be clearly seen from the observatory. Far above is the Milky Way, a starry smudge in the darkness. From his perch in front of a wall of windows lining the middle story of the 4-meter telescope, Richard Green, Kitt Peak's director, points northeast toward a feeble orange glimmer reaching into the night sky. That's Tucson, he says, about fifty miles away. International efforts to combat light pollution got their start in the city years ago. And while the urban population has grown rapidly, leaders have taken steps to help shield the city's glow. Then he points to the northwest, where the glow is a little fainter but covers a much larger swath of the horizon. That, he says, is Maricopa County.

"Once you're above the horizon, the contribution is about equal, and that's the shocking thing, given that Phoenix is about ninety miles away." Astronomers have rallied strong support for keeping the skies dark in Tucson and Flagstaff, which are both world-famous space science communities. Local governments in their counties, Pima and Coconino, have upheld lighting ordinances that protect the night skies by using lights sparingly and shielding them so they point down at the things they are meant to illuminate. But it is not the lights in those cities the astronomers worry about as much now. The Phoenix glow is what's growing.

It's edging north, too, toward Flagstaff. Since the observatory acquired the land, more has been revealed about topography and other characteristics that make a site ideal. Lowell Observatory's Anderson Mesa site—where most of its working telescopes are located—is not as perfect for astronomy as it once was, partly for those natural factors. But the creep of lights from Phoenix is certainly another one, and the Happy Jack site, where the Discovery Channel Telescope is set to go, is preferable partly because it's a little farther away.

Shying away from the light isn't a permanent solution. As cities grow, there are fewer great places to put telescopes. The future of observatories in Arizona, including Lowell, relies not just on running away from the glimmer but from controlling it, too, and people have made impressive strides on that front.

The International Dark-Sky Association began in Tucson with one man, who happened to be an amateur astronomer. He walked out of the University of Arizona medical center one night and was struck by the brightness of the lights in the parking lot. He complained, and found a willing helper in Al Tarcola, a long-time worker in the UA's facilities department. "As a youngster, I was up at Kitt Peak and helped build the observatories," Tarcola explained. "We started with the Health Sciences Center, changing the parking lot lights to low-pressure sodium." Back then, people were just beginning to voice concerns that they could not see the stars as well after dark. They called it light pollution. At a time when people were worried more about nuclear proliferation and air pollution, recalls John Polacheck, president of the Tucson-based Southern Section of the International Dark-Sky Association, "these people were on the fringe." Now, the association claims eleven thousand members in seventy-six countries and all fifty states.

There are three chapters in Arizona. Tucson's is active, holding monthly meetings and field trips to investigate various sources of lights such as the University of Arizona campus and sports fields. Flagstaff also boasts lively participation and support from the public and local governments. The city was named the world's first International Dark Sky City in 2001.

But participation has waxed and waned in Phoenix as dedicated members have come and gone. Scott Davis, an officer of the Tucson chapter, said that is partly because amateur astronomers are being chased out by the bright skies—these days, they are driving as far as Casa Grande, an hour southeast, to do any observing.

Former state Senator Gabrielle Giffords, a democrat from Tucson, sponsored astronomy-friendly legislation including the "good-lighting bill" that the governor signed into law in 2003. And she helped defeat legislation in 2004 that would have allowed for bright, blinking billboards along the state's highways. But she said she fought a hard battle with her colleagues from Phoenix. "Many members don't understand the importance of rural Arizona, don't understand the importance of industries in Flagstaff or Pima County, and they just don't care," she said. "They feel that Phoenix is the central driving economy to the rest of the state." And that feeling goes a long way when it comes to political haggling. Marc Spitzer, a former state senator who is now a member of the Arizona Corporation Commission, said he brought up the topic once during his tenure in the late 1990s. He had taken an astronomy course in college and sympathized with the scientists. But the opposition he met helped change his mind. He said demographics make restrictions a hard sell, and he worries about nighttime sports lighting coming under attack.

Dark skies proponents argue that astronomy is a vital industry that helps to define Tucson and Flagstaff. They say views of the stars are part of a high quality of life, that there are too few places that offer unencumbered views of the night sky and we shouldn't lose more. Over about two decades in Flagstaff, dark sky advocates have done battle with Arizona's major electricity provider, a major trucking company, Northern Arizona University, the Arizona Department of Transportation and gas stations trying to lure customers to their pumps with garish lights. Lowell Observatory's Wes Lockwood and the USNO's Chris Luginbuhl have led the charge. One successful longtime trend has been the replacement of mercury-vapor streetlights with astronomy-friendly, low-pressure sodium fixtures. The most recent challenge to Flagstaff's dark skies came in the summer of 2006, when the state's amateur softball lobby was vying for brighter lights at a new set of fields in Flagstaff. The dark sky advocates successfully argued for lights that were dimmer than the proposed ones but shielded, to direct the light toward the ground instead of the sky. They are half as bright as the leagues wanted, but brighter than the astronomers would have liked. Dark sky activists' other efforts have yielded similarly mixed results, where compromise has been the rule.

Flagstaff is a growing city with more than sixty thousand people. Some predictions have that figure rising to one hundred thousand people by 2050. Right now, Flagstaff puts out as much light as a typical city with forty-five thousand people. That's impressive, but it's not enough to guarantee the future of Lowell Observatory. Luginbuhl has said Flagstaff could be as bright as a city of thirty thousand if it reined in grandfathered and bootlegged lighting. Without continued diligence, lighting could go just as far in the other direction.

By building the Navy Prototype Optical Interferometer and the Discovery Channel Telescope, Lowell Observatory is doing its part to preserve its rich and long-standing legacy in astronomy. And so far, dark sky advocates have made great strides in keeping Arizona a good place for astronomers to work. But it's the next several decades that will reveal whether Arizona will choose bright lights—or a bright future for astronomy.

acknowledgement

The author wishes to thank the staff members at Lowell Observatory for their phenomenal generosity, warmth, and openness. Carolyn Shoemaker, Nat White, Wes Lockwood, Rose Houk, and Steele Wotkyns contributed their perspectives when this book was barely a concept. Bob Millis, the director, and Bill Putnam, the trustee, threw open the observatory's doors. Antoinette Beiser, the Lowell archivist and librarian, gave me a desk and brought me every little piece of paper I asked for. I think we even had fun. Henry Giclas, Otto Franz, Norm Thomas, Ted Bowell, Will Grundy, Mark Buie, Wes Lockwood, Dave Schleicher, Nat White, and Byron Smith showed wonderful patience in teaching me about their work. Henry Giclas's written reminiscences from his time at Lowell added color and depth to my understanding of the institution's stories. Brian Skiff and Kevin Schindler shared their impressively broad knowledge of the history of Lowell Observatory and astronomy in general. Claudine Randazzo, Eric Howard and Mike Frick at Northland Publishing were incredible to work with and provided boundless humor, latitude, and encouragement during the writing process. Without my parents, Dr. John Lawrence Minard and Christina Kochenderfer Minard, none of this, of course, could have happened. My dog, Shiva, and my cat, Kaya, have revealed themselves to be the most patient furred critters in the land.

Staff members at the American Institute of Physics' Neils Bohr Library were timely and gracious in responding to requests for access to their rare materials about Arthur Adel.

Support for this project from Discovery Channel Communications was in a class by itself.

Some of this material came from articles originally written for the *Arizona Daily Sun*, the *Arizona Daily Star*, and *National Geographic News*. Thanks to all of them for great experiences over the years and—in the case of the papers —copious use of their ink.

About the Author

Anne Minard is a freelance science writer and journalism instructor living in Flagstaff, Arizona.

She earned a bachelor's degree in biology from the University of North Carolina at Wilmington and a master's degree, also in biology, from Northern Arizona University in Flagstaff. She's written about science on the staffs of several newspapers and most recently as a freelance journalist.

As soon as Minard's reporting led her to Kitt Peak National Observatory, in southern Arizona, and into the domes of Flagstaff's Lowell Observatory, she became fascinated by the human quest to probe ever farther into the heavens.

Minard's writing has appeared in the *New York Times, Science* magazine, the *L. A. Times, Arizona Highways,* and numerous smaller publications. This is her first book.